THE SCIENCE OF SCIENTIFIC WRITING

THE SCIENCE OF SCIENTIFIC WRITING

JUDSON MONROE

English Department
University of California at Davis

CAROLE MEREDITH

Vegetable Crops Department
University of California at Davis

KATHLEEN FISHER

Genetics Department and Teaching Resources Center
University of California at Davis

KENDALL/HUNT PUBLISHING COMPANY

2460 Kerper Boulevard, Dubuque, Iowa 52001

Printed in the United States of America

401645 01

CONTENTS

PREFACE

Scientists ought to be good writers. They have the skills of analysis and organization which good writing requires, and they have the discipline and careful work habits which are also important. But many scientists dread writing; they simply don't know how to apply their skills to *this* particular task. This book is designed to solve this problem, to show scientists and students in the sciences how to apply the skills they have already acquired to their writing.

Regardless of the discipline, the methods described in this book **work.** We developed them, in part, by asking individuals who are both good scientists and good writers what they do when they put a manuscript together. We took "helpful hints" from dozens of such individuals and synthesized these into a coherent approach to language and writing. We also spent considerable time examining samples of scientific writing to determine how to distinguish the good from the bad.

Then we tested our methods in the classroom. The results were astounding. Nearly two hundred juniors, seniors, and graduate students in more than a dozen different fields took classes based on this approach. They wrote term papers, articles for scientific journals, and dissertations. In many cases, their grades improved. Some had articles accepted for publication. Quite a few prepared dissertations which their research committees accepted. And almost every student managed to reduce the number of drafts required to produce these good results.

To date, this book has been successfully used in such diverse areas as engineering, genetics, food science, psychology, veterinary medicine, physics, and vegetable crops. We have not found a scientific discipline in which the principles we describe do not apply, nor do we expect to.

The objectives of this book are, then, to show students in the sciences how to write well without spending weeks drafting and redrafting. These two objectives—*effectiveness* and *efficiency*—are achievable. Many students who had never written a good paper before were able to do so after brief instruction. Others used the book independently with similar success. Our emphasis is on data management and organization rather than on style or grammar because we believe these are the most critical and difficult aspects of scientific writing, and that once these are achieved, the remaining details follow easily.

This book would not have been possible without the contributions of dozens of faculty, staff, and students at the Davis campus of the University of California. We would like to thank the many students whose writing efforts provided us with numerous examples. We are also grateful to those colleagues who contributed ideas which helped us as we developed our scientific approach to writing and who reviewed our work as it progressed. Too many participated in this process for us to mention all, but we would like to give special thanks to Francisco Ayala, Bob Cello, Randy Curtis, Jon Fobes, Craig Giannini, Bob Keane, and Randy Schluter.

J. M.
C. M.
K. F.

Chapter One

THE SCIENCE OF SCIENTIFIC WRITING

THE SYSTEMATIC APPROACH

Early in their studies, scientists learn some basic approaches to analysis, to experimentation, and to general problem solving. They learn, in general, that they are more effective and efficient when they carry out their research in an orderly fashion. They learn to take things one step at a time, to build on others' work, to test their findings carefully before they go on to do subsequent research.

Scientific and technical writing can also be treated in this way, and many good scientific writers *have* developed a systematic approach to writing. When they write, they plan, draft, and edit in an orderly fashion. They test their work as they go along. They make sure that they've organized well before they sit down to put their ideas into words. They check their draft for clarity and accuracy before they proofread. As a consequence, their writing is consistently clear, precise, and effective.

Treating scientific writing in this way—as a science—is what this book is about. We have analyzed the writing process, breaking it down into a series of logical steps. Each step is designed to help the writer resolve one of the many planning, organizing, drafting, or editing problems which the scientific writer faces. The steps are presented in the order in which they ought to be taken. We also describe "tests" which the writer can and should use to evaluate his or her progress after each step. The tests help the writer to identify and correct problems that may be present at one step *before* he or she proceeds to the next step. In short, we've described the writing process in much the same way that we might describe an experimental process. Schematically, the procedure we suggest you follow looks like this:

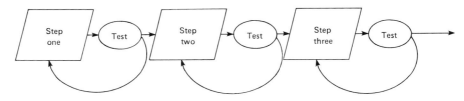

Figure 1.1. Step-test.

For this kind of orderly procedure to work, the writer must be able to analyze his or her writing **objectively,** must be able to judge the effectiveness of the paper in progress. Without this ability, the writer may finish a paper and discover that it must be done completely over again because of a basic error in organization or format. Good writers, like good researchers, have learned to avoid this problem; they never advance from one step to another until they've tested their previous work. The process of testing—of making sure you've got it right—ensures *real* efficiency. You never have to write six drafts of a paper if you do it right the first time.

All these procedures and tests are based on three simple concepts. An understanding of these principles will give you some insight into the rationale behind each part of the writing process.

UNDERLYING PRINCIPLES

The Relationship Between Thinking and Writing

There is an interesting "feedback" relationship between writing and thinking. Good thinking, of course, leads to good writing. Writing—the very process of putting words down on a page—promotes good thinking. This promotes good writing, which in turn further promotes good thinking. The positive feedback between these two processes is evident in a number of familiar situations. Students taking final examinations, for example, often find that their last paragraphs are much clearer and more interesting than their first. This happens because the process of formulating an idea on paper has sharpened their thoughts. Is it any wonder, then, that students frequently wish they could rewrite an examination?

The trouble with using the feedback process as the basis for writing is that it's inefficient. It involves a lot of drafts and redrafts, and no scientist has the time to spend in this kind of repetitive work. The writer must devise a way of taking advantage of the feedback relationship without all the hard work of writing five drafts. Many good scientific writers have done this, and their approaches have been synthesized here into a practical method for thinking/writing.

The Relationship Between Structure and Meaning

Many writers spend hours searching for a single word or phrase. Even when they find the right words, they still have trouble making their ideas clear. When this happens, it's usually because they have not considered that it is the *structure* of a thought which makes it clear or unclear. By structure, we mean simply the ordering of the elements of that thought. Take, for example, the simple thought: "X = Y." This idea can be structured in a number of different ways:

$$X = Y \qquad Y = X \qquad X - Y = 0 \qquad Y - X = 0.$$

2

There are other ways of ordering the elements of this idea, but we would merely like to point out that the easiest to understand is "X = Y" or "Y = X," because they are simplest in structure. The reader does not have to transpose terms from one side of the equation to the other. The reader does not have to know what "X" or "Y" represent, does not have to know the *words* at all. The relationship is still clear.

Applying this simple principle to verbal communication can yield some interesting results. Joseph Weisenbaum[1] developed a computer program, *Eliza,* capable of conversation of a very sophisticated nature. The program works in a remarkably simple fashion; it responds to human conversation on the basis of recognition of the kind of sentence structure the human uses. *Eliza* has an extremely limited vocabulary, but it can recognize a question structure, a command structure, or a cause-effect structure. It can respond to the idea the human is expressing by dissecting the structure of the sentence and extracting key words to insert into an appropriate reply sentence structure. Because *Eliza* is designed to focus on structure, it can even respond properly to sentences which contain some words from foreign languages. Most importantly, it isn't limited by its vocabulary. In short, *Eliza* communicates by structures; the program embodies a systematic extension of the reading-by-context which we all do at times.

If a computer can be programmed to communicate in this fashion, writers can also learn systematic approaches to such tasks as organizing, paragraphing, and phrasing. One of the primary goals in this book is to describe specific methods for each of these tasks; these methods are based on an analysis of the relationship between structure and meaning.

Reading and Writing as Conditioned Behaviors

Most writers feel that writing is a creative task, and a personal one. Many individuals tend to think of themselves as working in a vacuum, solving problems which are somehow unique. Perhaps they have this feeling because writing is a fairly complex process; each word represents some sort of decision on the writer's part. The likelihood that any two writers would choose the same thousand words to express an idea—and put them in the same order—is extremely remote. But this does not mean that the writer is not strongly influenced by *basic patterns* of expression which he or she may have learned over the years of schooling.

Schools of journalism long ago identified many of these writing patterns. Journalism students learn early in their studies that audiences will respond best when certain established paragraph and sentence patterns are used. As readers, we learn and relearn these patterns. We respond to them because we have been conditioned—much as we learn to say "Fine," when someone says "How are you?"

1. Joseph Weisenbaum, "Computational Linguistics," *Communications of the ACM,* August 1967, pp. 474-480.

The writer who is aware of the patterns which readers have learned can *use* this knowledge to help open channels of communication with readers. Such a writer can put new ideas into familiar contexts and can thus help the reader grasp concepts which might otherwise be difficult. This is particularly important in the sciences, and it may help explain why some scientists have little trouble writing clearly for a broad audience while others have difficulty communicating with anyone—even their colleagues. Some of the more useful and reliable patterns which a writer can adapt to scientific writing are described in detail at various stages in this book.

SUMMARY

This book attempts to treat scientific writing scientifically. It describes writing in a way that allows the scientist to gain control over this complex process. It provides procedures and standards that are based upon three simple concepts, (1) the feedback relationship between thinking and writing, (2) the influence of structure on meaning, and (3) the recognition of reading and writing as conditioned behaviors.

The book is organized chronologically. Each chapter deals with one step in the process of writing a scientific paper. Within each chapter, we have defined the problems that writers usually face at this particular stage. These problems are analyzed and solutions are suggested. In most cases, tests are also described that will permit writers to evaluate their work as it progresses. These tests provide a means of deciding, for example, whether or not a paper is organized well or whether it has actually achieved a goal such as "clarity." At the end of each chapter is a brief summary of the main ideas contained in it.

Since writing is a skill which requires some practice, exercises are provided at the end of each chapter. These do not require a great deal of time, and they will help you learn to utilize the methods in your own writing. Do this methodically. Use the skills of analysis which you have learned as a scientist, and apply these to your writing in an objective manner to evaluate your strengths and weaknesses. Using the guidelines in the text, make clear, conscious, and objective decisions about your writing.

Chapter Two

PLANNING FOR WRITING

THE PROBLEM

Writing is the process of turning vague thoughts into concrete paragraphs, sentences, and words on a page. This requires the writer to make clear decisions about what she or he wants to say. Because these decisions are interrelated—because one leads to or influences another—writing an idea down eliminates a lot of vagueness. Most writers don't begin this clarifying process early enough. They wait until they have finished all laboratory or field work, have collected masses of data, and are in a hurry to tell the world what they've discovered. At this stage, their heads are full of facts, analysis, opinions, methods, conclusions, implications, and questions about how to put all these together into a good paper. This is the worst time to begin writing; it's the time when making decisions is hardest because there are so many things to say. In addition, postponing writing delays the clarifying process that writing provides.

Because they wait until the last minute to think about writing, many writers find themselves writing draft after draft after draft in an effort to bring all their thoughts together. In this process, they quite frequently discover problems—bits of data which they didn't collect but which they *now* wish they *had,* questions which could have been considered earlier but which weren't. The writing process is thus complicated by frustration.

OBJECTIVES

The objective of this chapter is to describe an approach to writing which permits the writer to take advantage of the feedback between writing and thinking early in the research process. At the end of this chapter you should be able to write a prospectus that will serve as a useful guide for your research and writing, and you should be able to use incremental writing processes throughout your research.

THE PROSPECTUS

Defining the Question

Good scientific research begins with the process of defining the problem or question. This is a simple analytical process. It involves the researcher in separating a general problem into its components. The problem is thus expressed in terms of a number of parts that can be dealt with in quantifiable terms. For example, someone interested in public health might wish to examine the general question: "Why do some groups of people eat dirt?" This is a fairly complex question and, before the researcher can begin to consider it, it has to be broken down into parts. For example, the researcher might hypothesize that there are nutritional benefits to the practice of dirt-eating. This hypothesis could be further dissected into subhypotheses that (1) dirt-eating supplies minerals, (2) dirt-eating supplies calories, and/or (3) dirt-eating supplies bulk. These hypotheses are specific enough so that they can be studied; the researcher can collect data about them.

The end result of this dissection is that the researcher progresses from the "I'm thinking about doing some research on X" stage to the "I plan to study X subject and answer a, b, c, d, e, questions." Defining the problem in precise terms gives clear direction to the research. When this process is complete, the researcher is in a position to begin designing some means for collecting the data necessary to answer the questions that have been specified. The research can be conducted in an efficient manner.

This is a good time to *write* the first part of the paper—even though it must be done in a rather tentative form. In fact, writing should be an integral part of the research process. By doing so, you clarify and record your thoughts when they are fresh in your mind—before they get confused or forgotten or merely overwhelmed by the data you collect.

Points to Clarify

The Rationale for Your Research. Writing an introduction when your head is filled with findings and implications is extremely difficult, because introductions aren't supposed to present this kind of data. Why not, then, write down a clear statement of your rationale when you have an uncluttered view of it? You can use this statement to help you remember what you're doing as you get into the data collection stages of your research, and it can form the basis of the introduction to your paper. When you sit down to write, usually some simple modifications are all that is needed to transform the prospectus into an introduction.

The Specific Objectives of Your Research. Your reader will want these in the introduction. Make a list of them. Define carefully the kind of data you want to collect to meet each specific objective—to answer each question or test each hypothesis. If you can, define the units or terms which will be used to quantify this data.

The Scope of Your Research. Set down a list of key words or phrases that cover all possible subject-headings under which your research might be classified. This can be useful. It provides you with a guide for your library research; since many libraries have computer-search capabilities, a key-word list is essential. It also helps you to further delineate the scope and significance of your research. Finally, many journals will ask you to provide a list of key words.

Getting Feedback

Having written these things down, you may wish to show this material to a friend, a colleague, or to your professor. This will help you to see if your ideas are clear from someone else's point of view. The potential uses of this preliminary statement, or **prospectus,** about your proposed research are varied. In some cases, it is used in a very formal way, as a research proposal, to solicit funds from granting agencies. In other cases it is entirely for the benefit of the researcher/writer. Writing this material down at this stage will help you, and others, understand your research.

INCREMENTAL WRITING

The writing-clarifying process should not stop here. You don't have to put the rest of your paper off until the last minute. After you finish your literature survey, you can write your synopsis of this portion of research—while it's fresh in your mind. Doing this will also help you clarify your thoughts about what others have done and said about the subject. In writing this synopsis, you will force yourself to determine what data has and has not been collected. You'll then have a better idea of what data will have to be collected and how to analyze it.

Data Organization

Your next research step is one of the most difficult, to design methods for collecting and analyzing the data you need. Consider beginning these steps by setting up, in advance, the tables, charts, and/or graphs that you will use to present the data you have collected. This will give you considerable insight into where you're going. It also means that you won't have to do this at the last minute either. All you'll have to do when you get your data is to fill in these data shells! After you've done this, you should outline or preferably write a detailed description of the methods you choose for gathering data. Here again, you'll spend some time during the research stages in writing down descriptions of your methods while they're fresh in your mind.

The Benefits

The end result of this process, known as **incremental writing,** is that the writer has only two sections of a paper to write after all data is collected: the results and discussion. These contain the information that the writer is really interested in at the end of a research project; here the writer presents and discusses the findings and implications of the work. All the preliminaries, the introduction, the literature survey, and the methods, are substantially complete. They were written when *they* were important, when they required careful attention. Later, writing these sections becomes a mechanical exercise that tends to get in the way of the writer's thoughts about what he or she found out.

There are some side benefits to working on a paper in this way, to treating writing as a part of the research process. Not the least of these is a lowering of anxiety on the part of the writer. Most scientists are not professional writers. They gradually develop confidence as researchers, but they tend to be a bit hesitant when it comes to writing. Many students are even more than a bit hesitant; they're paralyzed by the very idea. Breaking the imposing task of writing a paper into discrete and brief segments helps many writers overcome this problem. It isn't, after all, that hard to write a few paragraphs—especially when that's all you have to write at one time. And so, a paper that might seem a momentous task if it had to be written in one step becomes a series of quite manageable small tasks.

The incremental process tends to reduce the number of drafts required to produce a finished paper, too. And with it, the writer-researcher has a simple means of getting advice at any stage in the research process by soliciting reviews of the written materials. Obtaining reactions to your work from a colleague (or professor) is helpful even when done simply through casual conversation; it is even more helpful when that colleague or professor has had an opportunity to read *written* descriptions of your research. People tend to treat something written more seriously than they do something brought up in conversation; you'll get more specific criticisms if you ask a colleague to review a written prospectus. Criticism is welcome. It can help you avoid mistakes both in the research itself and in drafting the paper describing it, mistakes which might otherwise prove troublesome. The incremental process can help you eliminate flaws in your work before they become public.

GETTING STARTED

The writing process begins with the *prospectus,* a simple statement of the rationale, objectives, and scope of your research. This provides the foundation both for your research and subsequently for a paper, thesis, book, or other research report. The following examples are given to acquaint you with some of the various forms that a prospectus may take.

Prospectus for a Library Research Paper

The theory that humans have descended from a primitive anthropoid line is

Rationale

commonly accepted as fact. Some recent paleontologists have suggested that evolutionary theory has not in any way been consistently supported by fossil evidence; the relationship of humans to the anthropoids, if any, has not been conclusively defined. ·

Objective

Can we actually account for the descent of the human race in evolutionary terms on the basis of current fossil evidence?

Scope

For this paper I will survey evolutionary paleontology to determine if we can currently prove from fossil evidence whether humans did evolve from a common anthropoid stem. Most arguments for evolution focus on: (1) morphological evidence, (2) taxonomic evidence, (3) tool use and tool making as evolutionary evidence, (4) changes in brain size, and (5) changes in speech and language.

This example gives the researcher/writer enough direction to begin the library research, although the list of areas to be examined is not a complete list of key words or subjects. Generating such a list would probably reveal to the researcher that the scope of the review is quite broad. A professor reading this as a proposal for a student research paper would certainly suggest that its scope be narrowed. But such a broad review article could be undertaken. If it were, the researcher's next step would be to further break the research task down into smaller parts. From these headings, the writer would generate an outline of the eventual paper. With this done, the library research could begin in earnest.

Prospectus for a Biological Research Project

Objective

In tomato, structural genes control the biosynthesis of the enzymes peroxidase-1, 2, 3, 4, 5; esterase-1; and phosphatase-1. There may also be regulatory genes involved. We can determine the extent of regulatory control on the biosynthesis of any one of these enzymes by comparing the enzyme levels of trisomic plants which carry three doses of the structural gene to normal diploid plants (carrying the usual two doses). If no regulatory control is involved, the extra gene in the trisomic plant should produce a proportional increase in the level of the enzyme. If no difference in enzyme level can be detected, then there are probably regulatory gene(s) involved in the

Rationale

Scope

biosynthesis of the enzyme. Various strains carrying mutations in the structural genes (isozyme mutants) will be tested. For each isozyme mutant, the following procedure will be followed:

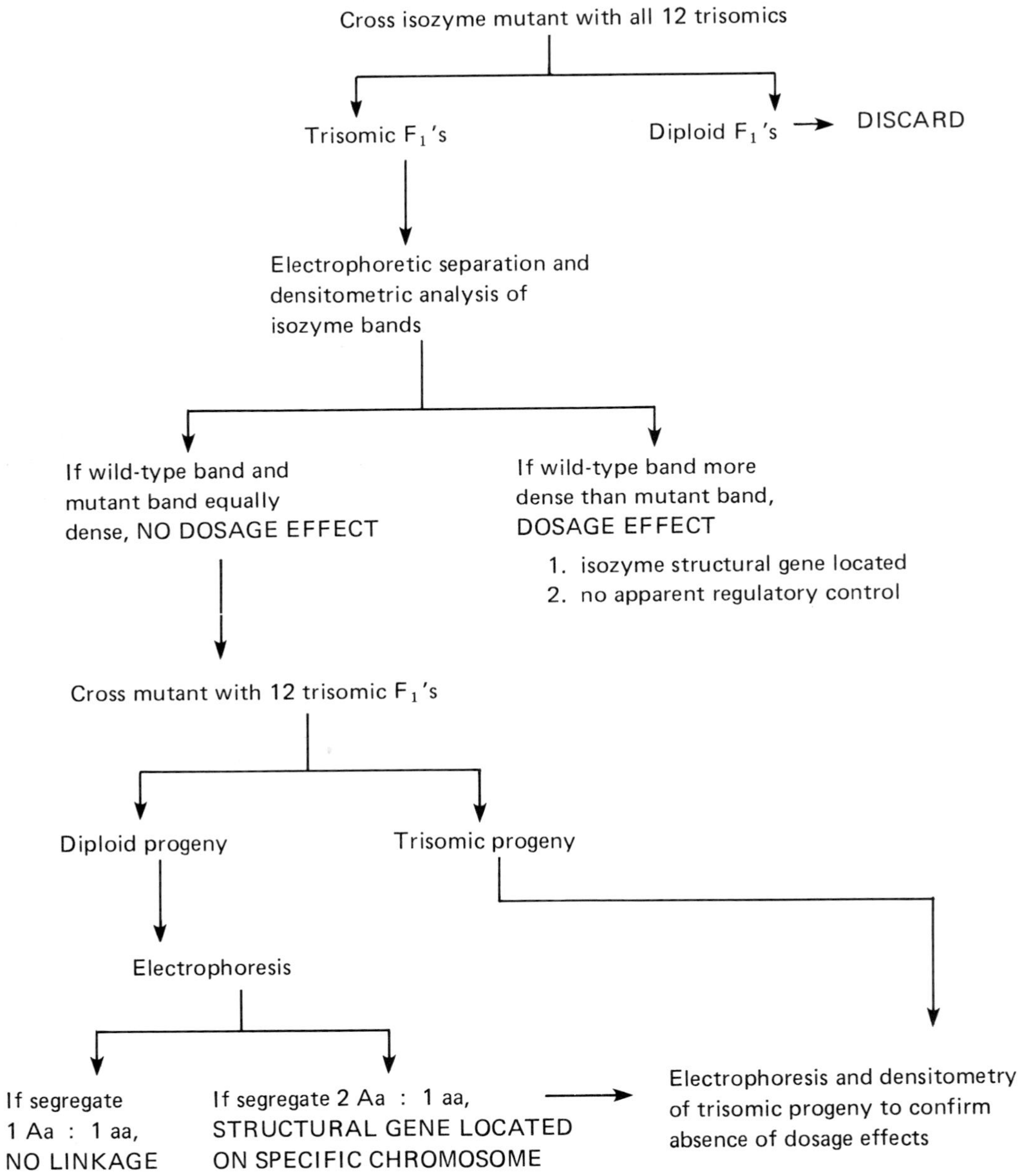

Cross isozyme mutant with all 12 trisomics
Trisomic F_1's
Diploid F_1's
DISCARD
Electrophoretic separation and densitometric analysis of isozyme bands
If wild-type band and mutant band equally dense, NO DOSAGE EFFECT
If wild-type band more dense than mutant band, DOSAGE EFFECT
1. isozyme structural gene located
2. no apparent regulatory control
Cross mutant with 12 trisomic F_1's
Diploid progeny
Trisomic progeny
Electrophoresis
If segregate 1 Aa : 1 aa, NO LINKAGE
If segregate 2 Aa : 1 aa, STRUCTURAL GENE LOCATED ON SPECIFIC CHROMOSOME
Electrophoresis and densitometry of trisomic progeny to confirm absence of dosage effects

Figure 2.1. Gene dosage flow chart.

The prospectus, especially one which isn't a final grant request, doesn't have to be in narrative form. Since it is a tool for the writer, a flow chart like this one may serve better than a long narrative description. It is easier for a reader to follow the proposed steps that the research will take in this form. At the same time, it provides a full "outline" of the project from which the writer can draft part of the final paper. Each step can be expanded and more fully explained in the final draft. The flowchart makes it almost impossible for the writer to get lost while carrying out research or writing.

A More Complete Prospectus

The following example has been excerpted from a seven-page prospectus written by a researcher *before* she began doing any of her research. It illustrates how much of a paper can be written even before the data is collected. The prospectus is written in past tense even though the work has not yet been performed so that it can be readily included in the final report. She has done a large part of her incremental writing at the outset, a common practice when the prospectus is being used as a grant proposal, and a wise practice even when the prospectus is intended only for the researcher (as this one is). As her experimental work proceeds, she will add to this, noting specific experimental details of her research methods and incorporating relevant new research reports that appear in the literature.

Propagation of Bitter Cherry, *Prunus emarginata*

INTRODUCTION

Prunus emarginata, a shrubby wild cherry, grows at moderately high altitudes in disturbed sites throughout the mountain ranges of the Northwest. It would be an excellent species for erosion control plantings because it occurs widely, has a spreading habit and an ability to survive at dry, nutrient-poor sites. The abundant white flowers in the spring, and red cherries which are eaten by birds in the fall, make it of further interest in shelter belts and wildlife conservation projects. But no one has successfully propagated this species.

LITERATURE REVIEW

Most of the standard treatments for dormant seeds have been tried on the seed of *P. emarginata.* The standard treatment to induce germination in most *Prunus* species is a cold moist stratification for 2 to 5 months, depending on the species (Hartmann and Kester, 1975). This treatment generally overcomes the dormant embryos common in the genus *Prunus.* Krier (1948) and King (1947) stratified lots of *P. emarginata* seeds at 1 °C or 5 °C for up to 24 weeks prior to planting at 23 °C. None of the seeds germinated.

Before the cold treatment can be effective, water must penetrate the endocarp which is generally stony in *Prunus,* but not impermeable. The endocarp on *P. emarginata* is extremely hard and suspected to be impermeable to water. To test the hypothesis of impermeability, King cracked the seeds and then stratified them for 4 to 24 weeks at 1 °C or 5 °C. The seeds stratified for 14 weeks at 1 °C had the highest germination percentage (5%). The study was repeated with sulfuric acid to reduce the thickness of the endocarp prior to stratification, but none of these seeds germinated.

Growth regulators such as thiorea can induce earlier or more complete germination of the seeds of some species (Hartmann and Kester, 1975). Krier and King soaked cracked and uncracked seeds in 0.5% and 0.25% thiorea for 2, 4, 8, and 14 hours. The seeds failed to germinate after stratification at 1 °C or 5 °C.

RATIONALE

Hartmann and Kester (1975) list four environmental factors important to the germination of seeds. These are water availability, temperature, light, and gas exchange between the embryo and the atmosphere. From what is known about propagation of the cultivated cherries, Mazzard (*Prunus avium*) and Mahaleb (*P. mahaleb*), the first two factors would seem to be the more important and were the ones considered for seed germination.

Prunus emarginata reproduces vegetatively, as is quite common among plants whose seeds show deep dormancy or those which grow under harsh and uncertain conditions. The factors which affect the regeneration of plants from cuttings are: the type of cutting material used, treatment of the cuttings, and environmental conditions during rooting (Hartmann and Kester, 1975). Three factors were considered for the selection of treatments: (1) the usefulness of the treatment for cultivated cherries, (2) the ecology of *P. emarginata,* and (3) the economic feasibility of the treatment. In erosion control projects a propagation technique must be economically feasible before it is considered to be truly successful.

PROPAGATION BY SEED GERMINATION:
OBJECTIVE AND SCOPE

A. Water Availability Tests. The cherries of *P. emarginata* ripen in late September, which is about the time that the winter rains begin in the Sierra Nevada. The similar Mahaleb cherry seeds are soaked for eight days prior to stratification. These factors suggest that an extended soaking period could be useful in the germination of *P. emarginata* seeds.

Four lots of 500 seeds (31.3g by weight) were soaked for 48 hours. After the first 24 hours the floating seeds were removed and counted to obtain the percentage of filled seeds. The remaining seeds were rinsed once and allowed to soak for another 24 hour period. They were then drained for stratification treatments. Four lots of 240 seeds were treated in a similar way except that they were soaked and rinsed for a total of eight days before they were drained for stratification treatments.

B. Temperature. Seeds of Mazzard cherries and some other *Prunus* species germinate best if they are stratified at a warm temperature (21 °C) for three or four weeks preceding cold stratification at 2 °C to 5 °C (Hartmann and Kester, 1975). A similar treatment is effective for the seeds of a large number of fall-ripening seeds with hard seed coats, where temperatures of 10 °C to 30 °C have been used. The warm period helps to soften the seed coat and improves water penetration. In some species the warm period is also important in after-ripening of the embryo.

A temperature of 2 °C to 7 ° is usually used for the cold stratification period, with a minimum of -5 °C and a maximum of 17 °C. The most effective temperature differs for different species or different stages of after-ripening. At the lower end of the temperature range no after-ripening can take place. At the upper end of the temperature range after-ripening not only cannot take place, but previously stratified seeds will revert to secondary dormancy. The germination percentage of such seeds will be reduced unless the seeds are germinated at a low temperature (3 °C to 7 °C for apples) or the seeds are subjected to another cold moist treatment. The temperature at which the seeds are stratified can be an important factor governing the germination of the seeds of many species.

"The temperature conditions most effective for germination may be similar to those in the natural environment . . ." (Hartmann and Kester, 1975). Soil temperatures have the most influence on the germination process but air temperatures (table 1) are all that are available for the altitudes at which *P. emarginata* grows in the Sierra (4,000 to 8,000 feet).

The above data plus exploratory experiments lead to the following stratification treatments: stratification at (1) 1 °C; (2) 6 °C; (3) 6 °C for one month, then 1 °C; (4) 15 °C for one month, then 1 °C. The lots of soaked seeds were randomly assigned to each of the stratification treatments. The seeds were stratified in polyethylene bags containing moist vermiculite until 50% of the seeds had cracked, or for five months, whichever was sooner.

The stratified seeds were sowed for germination in a randomized complete block design with two replications at 7 °C, two replications at 7 °C night and 13 °C day, and one as a control at a constant 21 °C. The germinated seeds were counted once a week for six weeks, or until no more appeared to be germinating. A germination percentage was calculated and an analysis of variance performed on the results.

The prospectus did not stop at this point, but this excerpt is long enough to illustrate how much of a paper can be prepared in rough-draft form before the research is even begun. The author prepared this some *months before she expected to obtain any of the results.*

In writing this much of her paper in advance, the author took full advantage of the feedback relationship between writing and thinking. She has set forth the background to her work and her basic methods, and in doing so has provided herself with written guidelines for her research. When it comes time to prepare the final draft for publication, she will have very little to write. She'll have to revise and refine this rough draft, but she will only have to actually *write* her results and discussion sections.

SUMMARY

To take advantage of the feedback relationship between writing and thinking, you should write in increments. Begin by writing a prospectus. Because it is written in the early stages of work, when the basic ideas for the research are not cluttered with details, the prospectus usually contains a particularly clear and lucid statement of the rationale, objectives, and scope of your research. The prospectus is, then, an excellent first draft of an introduction to your paper. When you're studying what others have done in your field, write your literature review. Before you actually begin your experiments, write out the procedures you'll follow. Once you determine how you will analyze your data, prepare the data shells and graphic shells that you anticipate using.

Doing this kind of incremental writing will improve your research by forcing you to put your ideas into concrete form. It will allow you to obtain feedback from your colleagues. And it will mean that when you've finished your research, you will be able to write your final paper quickly. Half of the paper will already be written, in rough draft, and the results and discussion sections will be pre-organized.

1. Below is a list of possible research topics. Select one, or make up your own, and
   (i) State the problem specifically;
   (ii) Break the problem into its component parts (i.e., specific questions which must be answered);
   (iii) Define the procedures necessary for answering each question.

TOPICS

1. obesity among American adolescents
2. glassware losses in university laboratories
3. time flies when you're having fun
4. Darwinian versus Lamarckian theories of evolution
5. the nutritional value of junk food
6. the role of petrochemical products in American life
7. the future of American wildlife
8. biological control of agricultural pests

Chapter Three

ORGANIZATION

THE PROBLEM

Having your thoughts clearly organized before you begin to write makes lucid writing possible. Few people would challenge this statement, yet poorly organized papers are sent to editors frequently. Why? The answer lies in the phrase "clearly organized." This is one of those vague terms which must be better defined before it can be studied or discussed. Just how do you organize clearly? How can you tell the difference between clear and unclear organization? How can you make sure that what you think is "organized" will be considered "organized" by your reader?

OBJECTIVES

The objective of this chapter is to describe a methodical approach to organization. Specifically, the chapter will cover (1) what organization is, (2) how to distinguish good organization from bad, and (3) how to go about organizing your work. At the end of the chapter you should be able to organize your thoughts about a subject in an orderly fashion and objectively evaluate the structure of your writing.

ORGANIZATION AND PURPOSE

Perception of Relationships

Organization is the grouping together of bits of information or data. It is an analytic/synthetic process which depends on the *organizer's* perception of similarities or differences among bits of data. For example, you might organize the data below in a number of different ways:

1, 5, 10, 12, 2, 6, 7, 143, 78, 356, 17, 21, 56, 13

If you perceive that some of these numbers are odd and some even, you might group them accordingly. Some are under 100 and some are over this number, so you might group them according to this criterion. Or you might leave them all together (they *are* all numbers) and organize them from smallest to largest or largest to smallest. The point is that your perception of the *relationships* among these bits of data would lead you to a particular grouping.

This perception of relationships is the key to organization; it is the process which distinguishes random from organized thinking and writing. The random thinker perceives no relationships and so cannot even begin to group segments of data together. But what is it that keeps thinking from being random? What is it that permits some people to perceive relationships quickly and leaves others struggling to find them?

Purpose Determines Structure

The answer is that organized thinking begins with *purposeful* thinking. A simple example will show how this is true.

When you walk into an office where a number of people are working, you perceive them performing various tasks. They walk about; they talk; they enter and leave their offices or office spaces. They seem unorganized, because the organization that does exist is not readily apparent. To understand how the office workers function as a group you must examine their relationships. You might ask how many different kinds of jobs are being performed and how many workers are doing each job. With this purpose the workers might be classified into groups such as *typists, bookkeepers,* and *file clerks.* If your purpose were to understand the interrelationships among the workers, the groups might consist of *managerial, staff,* and *clerical* categories. Each different *purpose* you generated for describing the organization of these office workers would also generate a particular grouping structure. **Organizing is, then, the process of creating structures for data, structures which are derived from the purpose of the study.**

Research and Writing Are Purposeful

If purpose determines basic organization, it follows that well-organized research and writing are purposeful. Whether this purpose is expressed in terms of a question to be answered, a problem to be solved, or an hypothesis to be tested, it is this purposefulness which accounts for the quickness with which some people organize their thoughts. As long as the purpose for collecting data is clear and precise, then the significance of each data bit is readily perceived. If the purpose is clouded or ambiguous, then the data also has little meaning.

One way to manage quantities of data is to design structures in advance to hold the data. It is helpful, for example, to design tables to put data into as it is gathered. As the information is collected, the table is filled up; the data is thus meaningful even as it is collected.

Suppose, for example, that you were interested in studying the responses of different strains of mice to testosterone, a male hormone. You could generate a number of different possible table shells, depending on your purpose for undertaking this study. If you were interested in determining whether the response to testosterone is genetically controlled, you might compare responses of several different strains of mice, using weight gain as an indicator. Your table might look like this.

TABLE 3.1

Mean Weight Gain (g) in Response to Increasing Testosterone Levels

Mouse Strain	μg Testosterone				
	0.1	0.5	1.0	10.0	20.0
HB1					
SV4					
R30					

If, on the other hand, you were simply interested in determining how weight gain and testosterone level are related in general, you could use the same data but you might want to average the results for all three strains.

TABLE 3.2

Mean Weight Gain (g) in Response to
Increasing Testosterone Levels

μg Testosterone	Weight Gain (g)
0.1	
0.5	
1.0	
10.0	
20.0	

Notice that the form, the visual structure, of the organizing device changes as soon as the purpose for organizing the data changes. It's the *purpose,* and *not* the data, then, which generates the structure.

This is a fairly simple principle, but it is significant because of its implications for both research and writing. For one thing, it means that the researcher can do much of the organizing of the final paper as soon as he or she has answered some very basic questions about the subject to be studied: Why am I conducting this research? What relationships do I wish to explore? What purpose will this data serve? These questions are frequently overlooked. Most people take them for granted, assuming that the answers are obvious. This is not usually the case. Conscious and deliberate consideration of these questions can help the writer begin the task of organizing early, before data begins to pour in.

GENERAL PURPOSE, GENERAL STRUCTURE

General Purposes for Organizing

Thinking of organization as a purpose-determined activity has other implications. Although there are an infinite number of *specific* purposes for grouping bits of information, there are a very limited number of general purposes. The most common of these are:

1. To define or describe
2. To compare
3. To find out why
4. To find out how
5. To find out what will happen if (or when) something is done

Each of these purposes generates a corresponding general structure. A comparison, for example, would look like this.

$$A - - B$$
$$A_1 - - B_1$$
$$A_2 - - B_2 \quad \text{or} \quad A, A_1, A_2, A_3$$
$$A_3 - - B_3 \quad \quad \quad B, B_1, B_2, B_3$$

Whenever a writer determines that her or his purpose is to compare, one of these basic structures will be used. This could represent the structure of a table or of the narrative. Characteristics of one thing being compared have to be ordered in such a way as to correspond with equivalent characteristics of the other thing(s) being discussed. Deviations from this basic pattern tend to obscure the comparisons. A reader faced with comparisons like the ones below would certainly be confused.

$$A - - B_3$$
$$A_3 - - B_1$$
$$A_1 - - B_2 \quad \text{or} \quad A, A_3, A_1, A_2$$
$$A_2 - - B \quad \quad \quad B_3, B_1, B_2, B$$

With this organization it is difficult to sort out the data and to comprehend its significance. The reader may give up after a short while, or may transpose the data to a more reasonable data shell in order to make sense out of it.

Well-organized writing is writing which allows the reader to discern the purpose of the writing and the way in which the data relates to that purpose. What the writer has to do when organizing is, then, to identify the *general purpose* of each section or subsection of the paper. For each general purpose, there is a corresponding *general structure* to be used. The writer then has to adapt this general structure to his or her specific purposes. A few of the commonly occurring general structures are described in the next pages. More will be mentioned in chapter five.

General Organizing Structures

Definition/Descriptive. The simplest expression of a definition is the equation "X = Y," where X is the item being defined and Y stands for the characteristic(s) which define or describe it. This equation defines the basic structure, or ordering of elements, for all definitions and descriptions no matter how complex. The subject, the thing(s) being defined, always comes first. The terms of the definition follow. Examine a dictionary, and you'll find thousands of examples of this general structure.

Most people don't have any trouble with simple definitions. They need to remember, though, that this pattern governs even more complex forms of definition.

Spatial/Descriptive. Consider the following description of the structure of acetic acid.

> Acetic acid is composed of two carbon atoms, two oxygen atoms, and four hydrogen atoms. The carbon atoms are joined by a single covalent bond. Three hydrogen atoms are covalently bonded to one of the carbons. A double covalent bond joins one oxygen atom to the other carbon, which also has a single covalent bond with the remaining hydroxyl group.

The same information might be more clearly presented in a picture or a diagram. The spatial relationships of the elements of acetic acid are drawn:

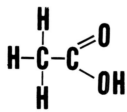

Figure 3.1. Acetic acid.

It's obvious that using a drawing or a picture to describe spatial relationships is a more efficient and effective way of transmitting this kind of information than is prose. Scientists have known this for years and use such visual short-hand as often as possible. But when it is not possible, spatial relationships can be genuinely troublesome to organize because, while a picture is simultaneous (you can see all elements at once or at least very nearly at once), writing isn't simultaneous. To describe spatial relationships, you have to give sequence to relationships which are not sequential.

Describing spatial organization in prose means that the writer must break an image into parts. There are a number of patterns which can be used to do this, depending on the kind of image being described. You can begin at the center of the structure and move out. You can start at the bottom and move to the top, or move from side to side. You can describe the major and minor features of the structure, and subdivide these even further. The important thing to remember is that, if you do use a narrative approach, once you establish a visual sequence you should be consistent and stick to it. To do this, it helps to use an actual picture or diagram of the structure you're trying to describe as a guide to your narration.

Comparison. The basic structure for comparisons has already been described. It is important that the characteristics of one thing being compared are ordered in such a way as to correspond to the equivalent characteristics of the other things being compared. The writer can either discuss all ''A'' data elements and then in the same order, all ''B'' elements, or can discuss corresponding ''A'' and ''B'' pairs, one at a time. Once a pattern is established, it should be adhered to carefully.

Cause-Effect. Cause-effect ideas are quite easy to organize because they are sequential, and the sequence can proceed in either direction depending on your goal. If your purpose is to examine the ramifications of a particular phenomenon, such as the draining of the Florida swamplands, the phenomenon (cause) is described first and its ramifications (effects) are described next.

When you are trying to determine what produced some event, the structure is reversed. The event (effect) is described first, and then the possible causes are considered. In this case discussing alternative causes first would bewilder the reader, because knowing what effect is being considered is essential in order to understand the data.

If there is a sequence of causes to be considered, state the observation to be explained, then present the causes for this occurrence *in sequence,* and restate the effect as the final part of the paragraph or section of the paper. Treat this kind of idea as you would a complex chemical reaction. State what product the process will yield, keep the steps in sequence from beginning to end, and give the product again.

How. Ideas that explain how something happens also ought to be organized sequentially. It often helps to begin your writing by sketching a flow chart or other visual representation of the process being described for your own use. This will make it easier for you to maintain a chronological order throughout your verbal description of the process.

If, Then or When, Then. These also are essentially chronological and ought to be set forth in clear chronological order. The "if" or the "when" is given first, followed by the "then." As with "how" ideas, simple diagrams can be useful tools to help you write clearly.

Using These General Structures

Knowing these simple structures can help you with your initial organization and provide a way to evaluate whether or not your writing is well-organized. The concepts can be applied at the sentence level as well as to paragraphs or whole sections of a paper. Suppose, for example, you had written the following:

> Among common garden beans, moths showed no egg-laying preference between golden wax beans and lima beans when presented with these and other plant hosts, each receiving insignificantly different numbers of eggs.

There are a number of problems with this sentence, not the least of which is that moths are not common garden beans. All of the problems result from faulty organization. This is a "when, then" idea. The information should be presented in chronological order.

> When moths were presented with golden wax beans, lima beans, and other bean varieties, they showed no egg-laying preference. Each plant variety received statistically equivalent numbers of eggs.

The information, arranged in chronological order, makes more sense and is easier to read.

The test for good organization, then, is really quite simple. First, determine what the *purpose* of the sentence, paragraph, or section of the paper is. Consider the *kind of ordering* appropriate to this purpose. Then, check to see that the writing is *consistent* with this pattern. There will probably be times when you'll decide to violate one or another of these patterns for a good reason, but that decision ought to be a conscious one.

HOW TO ORGANIZE A PAPER OR BOOK

A paper or a book may have one dominant structure, but it will necessarily also have numerous substructures (this book, for example, is chronologically organized to reflect the ordering of the writing process; within this structure, there are definitions, comparisons, cause/effect ideas, and other substructures). The writer must negotiate transitions, shifting from structure to structure. The shifts that occur from sentence to sentence are often quite difficult to describe, but the overall organizing process is fairly easy.

Suppose, for instance, you were writing a book about the ecology of a cultivated field. You would first have to decide on a general framework for the book. Since such an ecosystem would be severely disrupted at specific points in time, you might choose to organize chronologically by growing seasons. Your book would then have chapter titles like:

I. Precultivation
II. Plowing and planting
III. The growth period
IV. Mature crop
V. Harvest
VI. Post harvest

Within each chapter, you then ought to describe the characteristics of interest in a set order for comparative purposes. Each chapter might contain these sections:

A. Characteristics of the ecosystem
B. Dominant species
C. Other species
D. Population changes
E. Interrelationships of organisms

Each section could be further broken down into comparative subsections. Section "A," for example, might contain subsections describing weather, daylength, chemicals in the soil, nature of the soil, etc.

This process can go one or two steps further before it becomes too restricting. At each step, the writer must decide what the *purpose* of each section is and subdivide it accordingly. In this way, the paper or book can be carefully structured in advance of the writing. The framework can be developed even before the research is begun. This would certainly help the researcher see just what data needs to be collected at each stage. And the advantages to the writer are tremendous. He or she can concentrate on one bit of work at a time, moving rather easily from section to section without ever losing sight of the whole scheme of the work.

The primary advantages of this simple blueprinting method are that the writer can focus on a small segment of information and need be concerned with only one kind of organizing structure at a time. For example, while describing the subject or problem to be considered, the writer works with definition structures. Methods ("how" ideas) require a chronological structure. If the writer wants to explain why a particular method was chosen, there is a shift to cause-effect structure. The framework helps the writer keep track of what has preceded and what is to follow, and also helps prevent the use of inappropriate organizing structures.

This approach to organizing has benefits at the sentence level, too. Many grammatical errors are really nothing more than errors of organization, wherein elements of an idea are phrased in the wrong order. An awkward or difficult sentence can be analyzed by determining its purpose and then examining it to see whether the appropriate organizational structure has been used. By rewriting such sentences to correct problems of organization, phrasing and grammar problems are often also solved. Not all problem sentences can be corrected in this way, but many of the most serious can.

SUMMARY

When you sit down to organize your writing, follow this simple procedure:

1. Determine your overall purpose.
2. Determine what kind of structure is suitable for this overall purpose.
3. Define subpurposes.
4. Divide your paper into sections, each controlled by one of these subpurposes.
5. As you write, work from visual models wherever possible.
   a. If you're comparing, set up the comparisons in a table before you write about them.
   b. If you're describing something chronological or sequential, prepare a flowchart as a guide.
   c. If you're presenting an argument, prepare it in the form of an algebraic proof first.
   d. If you're describing spatial relationships, diagram them first.

If you approach writing in this way, you'll find that your first draft will be well-organized and that your ideas will be under control.

There is, of course, more than this involved in good writing. You often have to adjust your presentation to meet specific needs of your readers, and you have to consider style at a later point. This initial approach, however, will lead to a reasonably good draft on the first try. It focuses your attention on the things that are important at the outset. The other considerations can be given time later, as will be shown in subsequent chapters.

1. Below is a list of fifteen animals. Depending on your purpose, they can be organized in a number of different ways. Select two of the following purposes (or invent your own) and, for each, work out a chart or diagram showing how you would organize the animals.

    There is not necessarily a single correct solution. For a particular purpose, there may be several ways to organize this "data."

Data		
bat	elk	parakeet
cat	goldfish	pelican
chicken	horse	rabbit
dog	lion	trout
elephant	ostrich	turkey

Purpose for organizing:

size comparison
taxonomic classification
evolutionary relationships
classification of body covering
animals as a source of human nutrition
geographic distribution
animals as pets
domestication of animals

2. By now you should realize that visualizing the relationships among your data can guide you in your writing. Use the graphic representations below to test the organization of the paragraphs which follow. In each case, ask the following questions:

    (i) Are the relationships which you *see* consistent with those you read?

    (ii) Is anything out of order?

    (iii) Is anything left out?

    a. Compare this diagram and paragraph describing a program for testing the acidity of fresh-market tomatoes.

    Harvest ⟶ Shipping ⟶ Weighing ⟶ Storage ⟶ Testing

    Ten samples of fresh-market tomatoes were shipped from each of five major growing areas. Each sample consisted of twenty fruit and the fruit weighed approximately 160-180 grams each. The pH of individual tomatoes was measured and ranged from 4.2 to 4.5. The samples were shipped to the university by the extension agents in the various counties as they were harvested. Upon arrival, the tomatoes were stored at 20°C for six days before pH measurement.

b. Compare this diagram and paragraph illustrating the relationships between plants and animals.

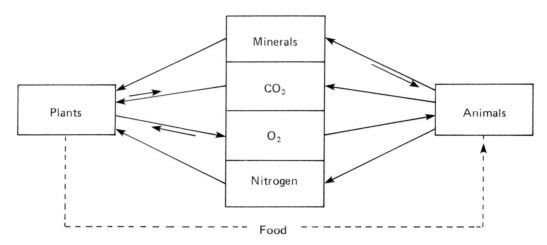

Figure 3.2. Relationships between plants and animals.

Plant and animal communities are closely related to one another. The main ways in which plants contribute to the maintenance of animal populations are to provide oxygen for animal respiration and to use the by-products of this respiration, thus maintaining an atmosphere favorable to animals. In addition, animals benefit from the plant community because plants provide a source of food. Even carnivores benefit from this; their prey are almost always plant eaters. The place of plants at the bottom of all food chains, no matter how complex, illustrates this relationship. In short, animals exist in an almost parasitic relationship with plants.

Chapter Four

GRAPHICS

THE PROBLEM

Words are extraordinarily useful tools, but they do have their limitations, as illustrated by the problems below.

1. Describe a double helix without using your hands.
2. Discuss the interactions of organisms in an ecosystem without drawing an arrow.
3. Explain the differences between the following sets of data points without drawing a graph:
$X_a = 1.2, 3.5, 2.4, 5.6, 6.2, 7.9$
$X_b = 1.3, 3.7, 4.8, 6.6, 8.4, 6.7, 4.3$

These are difficult tasks. The first borders on the impossible. This is because prose is basically a sequential medium. To write an idea you have to put one part down at a time. Some kinds of ideas just aren't suited to this sort of presentation. They are better suited to some form of graphic presentation which allows the writer to present an entire relationship in a single image rather than in a string of words.

The writer must decide when visualizations are important, when to put data into tables and figures. Once this is done, the problem becomes how to integrate the graphics with the narrative.

OBJECTIVES

This chapter describes guidelines for using and designing graphics, both as part of the writing process and as part of the finished paper. Using these guidelines, you should learn to use graphics in your writing, be able to decide when to include graphics in your papers, know how to integrate the graphics and the narrative effectively, and finally, you should find it easier to decide what kind of graphics to use.

THE USES OF GRAPHICS

Chapter Three introduces the idea that organization is the process of perceiving relationships. In some instances, such relationships are so complex that they are difficult for both writer and reader to understand until they are expressed in graphic form. The complex relationships which exist among the elements of a double helix are a good example. If you tried to describe a double helix with prose you would undoubtedly need page after page after page, and the reader would very likely find such a description impossible to follow. With just one picture, however, you can make yourself perfectly clear. Alternatively, you might write an equation to describe the curves. In any event, you abandon the narrative for the sake of clarity and efficiency.

Most scientific papers are a blend of narrative and graphic presentations of data and most writers recognize that graphics are an essential part of their papers. But graphics can be just as important to the writer as they are to the reader. They can provide the writer with guidelines for organization and for decision-making about how much discussion a paper requires.

Graphics function as a writing tool in two ways. First, they clarify the ideas which you are preparing to present to the reader. Second, they present data in a condensed form. Properly designed graphics often need very little explanation. If you design all the graphics for your papers before you begin to write, you can then minimize the amount of writing that you have to do. This can save you hours of drafting time as well as editing time. To see how this is true, examine the following paragraph:

> The effect of a special high calcium and protein diet on a group of twenty laboratory rats was studied. Twenty genetically identical rats were used as a control and fed a normal laboratory diet (XYZ Rat Chow). In the first week of growth, the special-diet rats gained an average of 1.3 grams each. The growth for the control group was an average of 0.8 grams. During the second week, growth rates expressed as weight for the two groups were: special diet average—1.6 grams, control average—1.1 grams. For the third week the growth rates were similar to those of the previous weeks. The special-diet group grained an average of 1.7 grams and the control group gained an average of 1.2 grams. At the month's end, this difference grew. The special-diet group gained an average of 2.6 grams during the fourth week whereas the control group averaged gains of 1.6 grams. The final measurements were taken a month later. During the second month of growth, the special-diet group averaged gains of 13.6 grams and the control group averaged gains of only 6.7 grams. Such performance indicates that the special-diet group benefitted from the special diet more as the experiment continued. There was considerably greater discrepancy between the growth *rates* of the two groups at the end of the experiment than there was at the beginning. This suggests that the influence of diet is cumulative.

This paragraph took about fifteen minutes to write, even though the data was clearly laid out in advance. The data was repetitive, and in order to avoid using the same sentence structure over and over, some fancy phrasing was required. But it's still boring and long-winded. In addition, it's difficult to extract from the paragraph exactly what did happen, even though an effort was made to summarize at the end of the passage.

A simple graph makes the whole process a lot easier for both reader and writer.

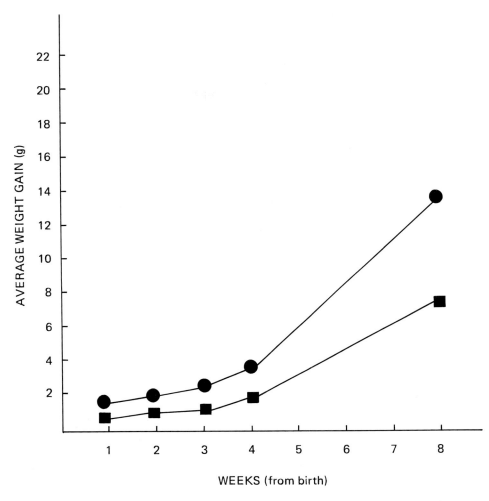

Figure 4.1. Average weight gain of laboratory rats fed high calcium and protein diet (●). Control group fed XYZ Rat Chow (■).

With this data already presented in graphic form, the accompanying discussion can be much simpler and more direct.

> Figure 1 shows the influence of a special high calcium and protein diet on weight gain in a group of twenty laboratory rats. The increasing difference between the experimental group and the control group suggests that dietary influences are cumulative.

Not only does the graph make the writer's job a lot easier, it also clarifies and visually summarizes the data. A reader can see at a glance that one diet produces better growth than the other. This is important because readers are increasingly becoming skimmers. They may read enough of a paper to understand its essential points and then concentrate on the data presented in graphic form. The data speak for themselves in this form and are more readily accessible.

You have many reasons, then, for preparing graphics before you actually begin to write the narrative sections of your paper. Indeed, many writers prepare graphics before they have collected any data at all. The process of designing these in advance, of preparing empty data shells for bits of data to go into as they are collected, helps them keep track of their research. In addition, it means that this part of the paper is done, in rough form, before the last minute rush to get something out. The incremental process works for graphics too.

WHAT KIND OF GRAPHICS TO USE

Deciding what kind of graphic would be most appropriate or most informative is not always easy. Many of the specific considerations will vary with the discipline and with the particular data set. There are, however, some general guidelines. **Tables** are used when it is important for the readers to see the exact values of the data, or when the data don't fit into any simple pattern. **Line graphs** are particularly useful for summarizing trends or interactions between two or more variables. They are used when the overall pattern, rather than the specific measurements, is important. **Bar graphs** are used to draw comparisons, particularly to dramatize differences. They are used more in popular writing than in scientific writing because they generally provide a very small amount of information while taking up a fairly large amount of space. **Diagrams** are useful for illustrating more complex relationships such as spatial configurations, pathways, processes, and interactions. **Flowcharts** are diagrams that are used for showing sequential processes. **Photographs and instrument tracings** are used when neither a verbal description nor a diagram would suffice in giving the reader the information that the writer wants to convey about an observable phenomenon.

In all cases, it is the *purpose* that determines the form of graphic finally chosen. Sometimes a writer may want to prepare several different kinds of graphics in order to decide which one is best suited for making a particular point. And the decisions don't stop there. Even when a table is the obvious choice, the writer must decide how to organize the data within the table. Or, if the choice is a line graph, should the coordinates be on a linear or logarithmic scale? All of these decisions should be made *before* the narrative is written. Your choice should be based primarily on the *purpose* of the paper, your reason for collecting the data in the first place. And it should have *integrity*. A line graph, for example, suggests a continuous relationship. Points should not be connected by a line if, in fact, no such relationship exists. Nor should a writer extrapolate beyond the data without making it clear to the reader what is being done. In all cases, remember that the text *follows* the graphics, and not vice versa.

HOW TO CONSTRUCT GRAPHICS

General Rules

Regardless of the type of graphic being constructed, it is useful to keep these general rules in mind.

1. The title should be **short, clear,** and **informative.** It should focus the reader's attention on the *purpose* of the table or figure. Long, complicated titles are to be avoided. You should not try to "tell all" in the title, and you should avoid repeating information already present in the graphic.

2. Each graphic should be able to stand alone as a **complete unit of information.** The title, legend, and labels should provide all the information the reader needs to understand the graphic. The graphics are constructed first and they should be independent of the text. The text is then built around them, and should be dependent upon the graphics.

3. The style chosen to present the data should **emphasize, not mask, the purpose** of the graphic. Clutter and nonessential information should be excluded. The effectiveness of the paper will be determined largely by the skills of the writer in selecting the best graphic style to emphasize each important point. Simply by reading over your graphic headings, you should be able to review the major topics covered in your paper.

Constructing Tables

Parts of a Table. Most tables have these main parts:

1. number and title
2. boxhead (which identifies the entries in the vertical columns)
3. stub (which identifies the entries in the horizontal lines)
4. boxhead for stub
5. field (which contains the data)

These are illustrated on the next page.

Title. The title announces the purpose of the table. It should be specific and brief. The titles of all tables for any one paper should form a coherent series. The title of the first table should provide all necessary orienting information which may then be omitted from later tables. For example, if one species is used throughout, it can be specified in the first table only. Or, a chemical can be identified specifically in the title of the first table, whereas an abbreviation or common name may be used in later tables.

Table 4.1. Pregnancy rate after transfer of embryos with different morphologic appearances.[1]

Number and title (label)

Boxhead for stub (label)
Boxhead (label)
Stub (label)
Field (label)

Elapsed time between recovery and transfer (hr)	Number pregnant/total recipients, by type of embryo transferred			
	Normal	Irregular	Abnormal	Total
4-5	3/6	. . .	. . .	3/6
5-6	2/3	. . .	. . .	2/3
6-7	1/3	. . .	. . .	1/3
7-8	4/4	. . .	0/4	4/8
8-9	4/5	. . .	. . .	4/5
9-10	0/2	3/3	0/1	3/6
10-11	. . .	1/1	. . .	1/1
11-12	. . .	. . .	0/1	0/1
12-13	. . .	. . .	1/1*	1/1
Total	14/23	4/4	1/7	19/34
(%)	(61)	(100)	(14)	(56)

*Two-cell embryo.

1. From Drost, M.; Anderson, G. B.; Cupps, P. T.; Horton, M. B.; Warner, P. V.; and Wright, R. W. Jr., 1975. *Journal of the American Veterinary Medical Association* 166(12): 1176-1179. Used by permission.

Example:
Table 1. Effect of soil temperature on growth of bush lupine.
Table 2. Effect of air temperature.
Table 3. Effect of salt spray.
Table 4. Effect of relative humidity.

Table titles should be consistent. Word order should not be varied unnecessarily. For example, the title of Table 4 might read, "Growth of bush lupine as influenced by relative humidity." The meaning is not changed, but the relationship of Table 4 to the other tables has been obscured by an unnecessary change in word order.

Format. Most journals have a standard format for tables. Therefore if you are writing a paper for publication, the requirements of the journal to which the paper is to be submitted must be considered. For example, a journal may have specific rules regarding the designation of column headings or units of measure. Similarly, guidelines are usually provided for theses and books. Becoming familiar with the requirements of the journal or publisher before preparing tables may save the author considerable time.

Grouping Data. The data presented in a table should be grouped logically. Control values are generally given first. Columns or rows which will be frequently compared should be adjacent. If a pattern has been set up in a previous table, it should be retained in subsequent tables when possible. Natural gradations in the data may be stressed by appropriate arrangement of the rows and columns.

Shape. The shape of a table is determined not only by the data, but also by the format of the journal in which the article is to appear or by the format of the book or report. The shape of a table can often be altered considerably by rotating the table 90 degrees. This change may result in a table of more pleasing appearance and heading words may better fit their allotted spaces.

Boxhead and Stub. The boxhead and stub should be precise and concise. They should be as specific as possible, and easy to understand.

Footnotes. If more than a few words are needed in order to fully identify a column or row, additional information can be placed in a footnote, usually indicated by a superscript number, letter, or symbol. Footnotes are also used to supplement the title and field entries in a table. The footnotes should appear immediately below the table.

Field. The data contained in the field should be internally consistent. For example, if the first number in a column is carried to three decimal places, the following numbers should be too.

Efficiency. Space is precious in a table. Every value which is not essential to the purpose of the table should be eliminated. Repetitious words and reference values can often be eliminated or transferred to a footnote. Not only is an uncluttered table easier for the reader to follow, but it is also more likely to be accepted by journal editors and thesis committees. It is also wise to avoid repeating a substantial amount of the boxhead and stub information in two different tables in one paper. More often than not, when this occurs, you will find that you can readily condense the two tables into one.

Constructing Graphs

Parts of a Graph. Graphs consist of:

1. number and legend
2. axes (usually 2, sometimes 3)
3. field

An example of a line graph is shown below.

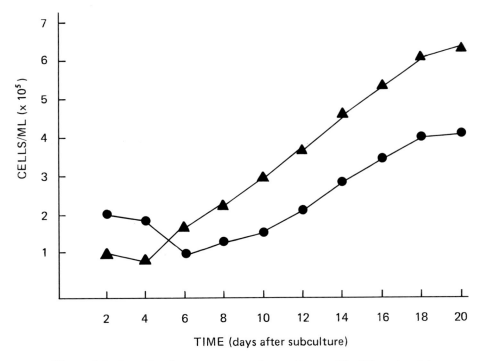

Figure 4.2. Growth of tomato suspension cultures with different nutrient media. CS5 medium (▲), PSC medium (●). Each point represents the mean of 4 measurements.

Legend. A legend contains the title and identifies the symbols used in the graph. It may also point out important features of the graph and provide experimental details. The legend should be clear and concise enough that the reader does not have to refer to the text to understand the graph. Legends for similar graphs should be consistent in sentence structure.

Axes. Be sure to label each axis clearly, indicating both the variable and the units of measurement. Mark the intervals on both axes.

Field. Keep the field uncluttered. Combine several curves on one graph when possible, but do not exceed three or four curves per graph (if the lines cross) or at most five or six (if the lines are distinct and nonoverlapping). Use simple symbols (closed circles, open circles, etc.) to indicate data points. Some journals and publishers will require that you use specific symbols.

Be consistent. If you have several graphs showing attributes of one set of things, use the same symbols throughout the series. In general, it's a good idea to show all data points along the graph. When each data point is the mean of several measurements, you may want to indicate the standard deviation or standard error by a vertical bar at each point.

Never extrapolate a curve beyond the points observed (unless you explain clearly that you are extrapolating). Do not draw a curve between any data points without considering the implied continuum.

SUMMARY

The primary function of graphics is to help the writer understand and express relationships which are either too difficult or too cumbersome to express in narrative form. Graphics are best designed before any narrative is written because they help clarify the ideas to be discussed and help the writer determine how much discussion is really needed.

When designing graphics, the writer should consider the purpose of the visualization and choose an appropriate format for the data which will be collected, placed within the "data shells," and presented to the reader. Whatever form you choose, make sure that the graphics you design illustrate *accurately* and *clearly* the findings you are presenting.

1. Summarize the data given below in (1) a table, (2) a line graph, (3) a diagram and (4), a bar graph. Construct each graphic in complete form, with title and legend. Label accurately.

   The data is derived from a food science experiment. The purpose of the study is to determine the influence of temperature on the respiration rate of vegetables in storage. The results are expressed in milliliters of carbon dioxide produced per kilogram of vegetable per hour. Measurements were taken at $0°$, $10°$, $20°$, and $30°$ Centigrade. These are the results.

   Tomatoes: at $0°$—5 mls/kg-hr, at $10°$—8 mls/kg-hr,
                 at $20°$—20 mls/kg-hr, at $30°$—37 mls/kg-hr.
   Broccoli: at $0°$—11 mls/kg-hr, at $10°$—22 mls/kg-hr,
                 at $20°$—89 mls/kg-hr, at $30°$—136 mls/kg- hr.
   Onions: at $0°$—4 mls/kg-hr, at $10°$—5 mls/kg-hr,
                 at $20°$—6 mls/kg-hr, at $30°$—9 mls/kg-hr.
   Lettuce: at $0°$—4 mls/kg-hr, at $10°$—12 mls/kg-hr,
                 at $20°$—23 mls/kg-hr, at $30°$—42 mls/kg-hr.
   Potatoes: at $0°$—4 mls/kg-hr, at $10°$—5 mls/kg-hr,
                 at $20°$—6 mls/kg-hr, at $30°$—29 mls/kg-hr.

   What conclusion(s) can you draw from the data? Are they reflected in your titles? Which graphic provides the most effective summary? Why?

Chapter Five

ANTICIPATING
READER RESPONSE

THE PROBLEM

Having organized your data, you now have a firm command of the ideas you want to present to the reader. Your next task is to decide how to present these ideas most effectively. Your object as a writer is to make the reader's job as simple as possible. *You* don't appreciate a writer who makes it difficult for you to understand something; you want to avoid doing this yourself. The problem you face is, of course, that you don't have the reader sitting next to you while you write. You can't ask the reader if a paragraph makes sense or not. You have to anticipate the reader's reaction to what you have written. You have to predict what questions the reader will have, and answer them in advance.

OBJECTIVES

This chapter describes some ways you can predict reader responses to the manner in which you have presented certain kinds of information. At the end of the chapter, you should be able to test your writing so as to be sure that you're giving the reader the right information in the right order.

CONDITIONED RESPONSES

The successful writer is like the successful comedian, the person who can tell a single joke and make a million people laugh. Successful writers are those who can induce a large number of different readers to respond positively to (or at least to understand) their writing. Both writer and comedian have the same task: to design communication which will elicit a predictable response from a diverse audience. This task is made easier because people's responses are *conditioned*.

Conditioning takes place throughout the lives of all writers and readers. Consider for a moment that writing is like talking, except that there is no one there to reply. You know, though, that when you talk with someone, you get predictable responses to your statements or questions. For example, when you say something like "I've got a problem," your friends respond with a simple question "What happened?" or "Why?" In the same way, if you ask a question, you'll usually get a response in the form of an answer, often the familiar "I don't know."

The specific content of such stimulus-response exchanges in conversation can vary almost infinitely. But the *general* stimulus-response *patterns* will not vary. Suppose, for example, that you wrote the following sentence:

> Hemolytic diseases occur naturally in newborn horses when a dam produces antibodies against her foal's erythrocytic antigens.

This sentence introduces a subject and thus generates some predictable expectations on the part of the reader. First, the reader may wish a definition of "hemolytic" and "erythrocytic." Even the reader who is familiar with such terms may wish them defined. In a general way, readers would also like to now know some more detail about this problem. How often does it occur? How serious a problem is it? What happens when this occurs? Knowing that these are questions the reader will ask about any statement of this sort, you can prepare answers in advance.

As you write, you generate responses with every sentence and word you put down on paper. You make a statement and the reader then will expect you to follow up. Your task, then, is to make sure that you don't generate false expectations. If you lead the reader to expect something, you must make sure that you provide him or her with this expected information. When you establish a stimulus-response pattern, you should do so consciously. Be aware of the effects your writing will have on your readers.

With this rule in mind, examine the following abstract. Did the writer satisfy the expectations which he or she generated?

FLAVOR BASE COMPOUNDING FOR FOOD PRODUCTS: IMPORTANCE AND APPLICATIONS

The study of flavor ingredients in food science and technology has been developing steadily with the growth of the food industry and people's desire to render food more flavorful. The worldwide increase in food demand is leading experts to predict a food crisis in the near future. It is necessary to obtain new sources of food which will meet with general acceptance. This acceptance depends upon flavor as well as nutrient content. Creation of flavors is an art as well as a science; the flavorist must handle over two thousand flavor ingredients whose odors complement each other and create new and unique flavors. Further, competence in a large number of fields is required to encourage the creation of potentially acceptable flavors: organic chemistry is needed for an understanding of the chemical nature of flavor ingredients; analytical chemistry for the identification of flavor constituents; statistics for the evaluation of flavors; etc. Even the preparation of simple plum flavor requires at least a basic understanding of all the above fields.

The author of this abstract has generated two expectations in the title of the paper which aren't fulfilled at all. First, the title suggests that the discussion will include a section about "applications." It doesn't. Instead, the writer presents a discussion of "problems." The second expectation is more specific; the reader expects the abstract to begin with a discussion of the "importance" of flavor base compounding. This subject isn't introduced until the second sentence, and then it isn't introduced very clearly.

These problems don't make it impossible to figure out what the author is getting at, but they often leave a reader rather uneasy. They can easily be avoided once they have been identified. To identify these problems, the writer must be aware of the stimulus-response patterns he or she is generating with a title or first sentence. Once such a pattern is recognized, it's easy to organize the rest of the paragraph to fit he pattern.

FLAVOR BASE COMPOUNDING FOR FOOD PRODUCTS: IMPORTANCE AND APPLICATIONS PROBLEMS

The worldwide increase in food demand is leading experts to predict a food crisis in the near future. To avert this, new sources of food must be developed and accepted. This acceptance depends on flavor as well as nutrient content. Thus, the study of flavor ingredients in food science and technology has been developing steadily with the growth of the food industry and people's desire to render new foods more flavorful.

Creating flavors is an art as well as a science; the flavorist must handle over two thousand ingredients whose odors complement each other and create new and unique flavors. Further, competence in a large number of fields is required to create potentially acceptable flavors: organic chemistry for understanding the chemical nature of flavor ingredients; analytical chemistry for the identification of flavor constituents; statistics for the evaluation of flavors; etc. Even the preparation of simple plum flavor requires a basic understanding of all these fields.

STIMULUS-RESPONSE PATTERNS

Communication would be incredibly difficult were it not for a number of standard stimulus-response patterns which are used over and over again in writing and speaking. These patterns enable people to communicate with relative speed. There are dozens of these basic patterns; some of the more common are:

Question-Answer. When you generate a question in writing, the reader will expect you to answer the question—soon.

Problem-Solution. If you present a problem, the reader will expect a solution or an explanation of why no solution is forthcoming.

Cause-Effect, Effect-Cause. Whether you have mentioned a cause first or an effect first, once you have mentioned one, the reader will surely expect you to mention the other.

General-Specific. When you make a general statement, the reader will expect to be supplied with specifics which clarify, qualify, or explain the general statement.

Description-Analysis. The reader will generally expect you to discuss or analyze anything you take the time to describe. This includes, of course, any data you introduce. Readers presume that you have a reason for introducing data or describing something. You must explain that reason.

Analysis-Summary. As analysis is often complicated, readers have come to expect that the writer will tie things together with a summary.

Summary-Conclusion. Readers expect the writer to have thought about the information presented and to have arrived at some kind of conclusion. They expect the writer to end a paper with an answer to that important question: "So what?"

These are a few of the most common stimulus-response patterns which you ought to keep in mind as a writer. The first three are also general organizing structures. The last four are particularly important because they serve as both general organizing structures for papers or sections of a paper, and as a general format for individual paragraph structure as well. They are described in more detail below.

A GENERAL FORMAT

Since reading is a conditioned behavior, it is not surprising that readers have come to expect to see certain overall patterns in what they read. Journalists have long been aware of this and have reinforced this expectation by using certain fairly rigid patterns for presenting news and entertainment articles. In the same way, scientists have adopted general patterns for articles. These have become so standard that readers expect them without being conscious of them. As a scientific writer, then, you can expect that your readers will want the following kinds of information in the following order:

Establishing Statement. A reader needs and wants to have some context established. At the beginning of each major paragraph or section, the reader expects a statement of purpose, a topic statement, a problem statement, or some pertinent background information. Which of these alternatives is used will depend on the subject and purpose, but in any case, the first one or two sentences should function to establish context.

Detail: Descriptive, Elaborative, or Qualifying Statements. Once you set the context of your discussion, you need to give the reader detailed information. Answer the questions: Who?, What?, Why?, or Where? Don't answer all these questions every time, just the ones that are important to your discussion.

Analysis. Once you've given the reader some information, you can analyze the relationships in the data. Here's where you explain things to the reader.

Summary. Analysis is usually fairly complex and readers usually need a brief summary statement.

Conclusions. Here's where you answer the question, "So what?" Writing is a purposeful activity, and in scientific writing, the purpose usually points toward some very specific conclusions. The writer leads the reader to those conclusions through the logical sequence of steps given above. The reader expects the conclusions to be explicitly stated at the end of the sequence. If you've summarized a point but do not wish to present conclusions until you've made another point, the conclusion is replaced with a transition statement.

Using the General Format

This simple pattern, which is actually quite close to the overall pattern of most scientific journal articles, gives writers a means of effectively testing their writing. Since it is a pattern that a reader will generally expect to find in a paragraph, a section of a paper, and in an entire paper, the writer can use it as a guide for making decisions about the ordering of information. It's just a guide, since not all paragraphs or sections will contain every one of these components in this order, but it can be helpful.

This basic paragraph pattern can be used easily to determine whether or not the information you have to present to the reader is likely to be easy for the reader to follow. After you've drafted, analyze your writing by marking each sentence according to the function it serves—establishing statement, detail, analysis, summary, or conclusion. If you find that you're presenting information in this order, you can be fairly sure that your readers will be able to understand your writing.

The abstract which follows is a good example of writing which adheres to the basic paragraph format. We've separated the functional sections with slash marks (/) and labeled each section in the margin.

Crack propagation in a typical structural ceramic (porcelain) is accompanied by acoustic emission./	Establishing statement
Two types of emission are detected./ The first type is caused by slow growth of the fracture-initiating flaw; the emission rate depends primarily on crack velocity./	Establishing statement Detail
Failure prediction using this source of emission can be effective, however, only if low-level emission, which may be related exclusively to crack growth, can be detected./	Analysis

The second source of emission, which occurs during bulk stressing, is the cracking associated with second-phase particles (quartz particles in porcelain) as a result of the combined action of the applied stress and local thermal and mechanical stresses./	Detail
An analysis for predicting emission rates is developed and forms the basis for using this source of acoustic emission in failure prediction.[1]	Analysis Summary

This abstract, from an award-winning paper, demonstrates both the usefulness of the basic format and the need for the writer to be flexible in using it. The paragraph begins with simple establishing information; the writers have taken their responsibility to explain the foundation of their research seriously. Even a layman can understand these two sentences. Then the authors break the detail and analysis sections into two parts, leaving until last the part with which they are primarily concerned. This break is necessary; without it the reader might get the two processes mixed up. Within each subsection, the authors are careful to describe before they begin to analyze. Their summary, while brief, is sufficient for an abstract.

Trouble-Shooting

An analysis of your writing will sometimes reveal problems. Even when you're tried to organize carefully, the writing process itself can get a bit chaotic. When this happens, writers often get a vague feeling that something's wrong. Marking each sentence according to its function can be especially beneficial when this happens because it helps you identify the specific problems which you may have created in the process of getting your ideas down quickly. Once you know exactly what's wrong, you can remedy it quickly and confidently.

The following abstract is disorganized. The original author recognized this, but couldn't figure out what to do. Marking each sentence according to function gave her a clear idea of how to solve the problem.

CONSUMER REACTION TO
CARBONATED BEVERAGE
PACKAGING

Safety, ecology, convenience, economy and ingredients are important factors in the decision to purchase a carbonated beverage./	Analysis
For those reasons and others, "Anpac" container introduction is likely to be controversial./	Summary and Conclusions

1. From Evans, A. G. and Lindzer, M. 1973. Failure prediction in structural ceramics using acoustic emission detection. *American Ceramic Society Journal* 56(11): 575-581. Used by permission.

The Monto Can Company and the Best-Brew Company are among several manufacturers who have been developing and planning to market the lightweight, strong, transparent packages. Best-Brew in Monto Can's "Anpac" containers will be test-marketed soon.

Detail

A Food Additive Order, issued February 12, 1975, by the Food and Drug Administration, cleared acrylonitrile/styrene copolymer as a component of packaging materials in contact with nonalcoholic beverages./

Resins of acrylonitrile/styrene copolymers (60-85% AN content by weight) possess excellent carbon dioxide, oxygen, and moisture barrier properties which render them suitable for molding into carbonated beverage containers./ Bottles made from these high nitrile materials retain carbonation and fill-line levels throughout expected market shelf life./

Establishing statement

Detail

Two types of market testing techniques, consumer focus groups and an attitude survey, indicate consumers have mixed feelings about the introduction of plastic carbonated beverage containers.

Analysis

When you read through the abstract, you realize that the information is not presented logically. Yet efforts to improve it may result in five or six more rewrites if the "stab in the dark" approach is used. The problems can be easily corrected in one step, however, if you use a *systematic approach.* By examining each sentence to determine what kind of information it contains (establishing, detail, analysis, summary, or conclusion) and jotting the type in the margin, your task is simplified. You simply rearrange the parts to fit the pattern, as shown below.

CONSUMER REACTION TO CARBONATED BEVERAGE PACKAGING

Resins of acrylonitrile/styrene copolymers (60-85% AN content by weight) possess excellent carbon dioxide, oxygen, and moisture barrier properties which render them suitable for molding into carbonated beverage containers./ Bottles made from these high nitrile materials retain carbonation and fill-line levels throughout expected market shelf life.

Establishing statement

A Food Additive Order, issued February 12, 1975, by the Food and Drug Administration, cleared acrylonitrile/styrene copolymer as a component of packaging materials in contact with nonalcoholic beverages.

Detail

The Monto Can Company and the Best-Brew Company are among several manufacturers who have been developing and planning to market the lightweight, strong, transparent packages. Best-Brew in Monto Can's "Anpac" containers will be test-marketed soon./

Two types of market testing techniques, consumer focus groups and an attitude survey, indicate consumers have mixed feelings about the introduction of plastic carbonated beverage containers. Safety, ecology, convenience, economy, and ingredients are important factors in the decision to purchase a carbonated beverage./ For those reasons and others, "Anpac" container introduction is likely to be controversial.

Detail

Analysis

Summary and Conclusions

This kind of rearranging can often be done without major changes in sentence structure or phrasing, and it almost always leads to paragraphs and sections of a paper which are easier for the reader to understand. All it requires is that you think carefully about the function of each sentence in a paragraph. Once you decide what the sentence *does,* you can quickly place it in proper relation to the other sentences in the paragraph.

SUMMARY

Good writing results when a writer anticipates the reader's responses. A basic stimulus-response pattern is described which can be utilized within any organizing structure.

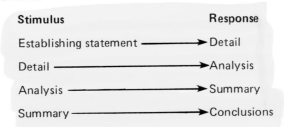

Stimulus	Response
Establishing statement ⟶	Detail
Detail ⟶	Analysis
Analysis ⟶	Summary
Summary ⟶	Conclusions

The general format generated by this basic stimulus-response pattern can be used first to help you control information as you draft your paper, and second, to diagnose and correct your writing when you've completed a draft.

The format provides guidelines, not fixed rules. The writer needs to be flexible and to vary the format as needed to fit specific circumstances. Following the format most of the time, however, and particularly when you are in doubt about doing otherwise, will help you produce consistently clear writing.

1. For each of the following statements, describe your response as a reader and determine what kind of statement(s) might follow to meet that response. (What questions has the writer raised in your mind? How should they be answered?)

   a. Much progress has recently made toward the understanding of subatomic particles.
   b. Although the tobacco hornworm (*Manduca sexta*) and the tomato hornworm (*Manduca quinquemaculata*) are almost identical, there are subtle morphological differences between the two species.
   c. The use of diethyl stilbestrol has recently been questioned.
   d. A novel method for fixation of lungs at inflation volume preserved particles and cells on the surfaces of airways from trachea to terminal bronchioles.
   e. Growth hormone administered to pregnant rats is said to enhance brain development by the time of birth or in early life.

2. Below are two sets of statements. Each set needs to be organized. In each case, rearrange the sentences to produce a clear paragraph (or paragraphs) according to the basic paragraph format described in this chapter. Determine the function of each statement. Is it establishing information, detail, analysis, summary or conclusion? You may change some of the wording or combine statements as necessary to produce an organized flow of information.

   Set I

   a. Schulz and Jensen (1968) have shown that the highly polar *Capsella* egg becomes even more polar when fertilized and that the polarity determines the plane of the first cell division.
   b. Cell polarity is especially evident in eggs and zygotes.
   c. Torrey and Galun (1970) have demonstrated the importance of cell polarity in *Fucus*.
   d. The presence of a strong chemical or electrical gradient across a cell is called cell polarity.
   e. Cell polarity is undoubtedly of critical importance in differentiation.
   f. When the polarity of a *Fucus* zygote is disturbed by a high sucrose concentration, abnormal embryos develop.

   Set II

   a. Cavernicoles (cave-dwellers) have altered endocrine systems.
   b. In the absence of light, endocrine regulation cannot occur.

c. There are two primary sources of food energy in the cave, both of which originate on the surface.

d. In epigeans (surface-dwellers), increasing daylength often stimulates reproductive activity.

e. One of the most important characteristics of the cave environment is the absence of light.

f. Strictly phytophageous (plant-eating) animals are excluded from the subterranean world.

g. In epigeans, light plays an important role in the regulation of the endocrine system.

h. Caves are devoid of green plants, as photosynthesis cannot take place in darkness.

i. Cave animals must turn to food sources other than green plants.

3. Evaluate the following passages to determine if the author has given the reader sufficient information and if this information is in the right order. Read the passages through once to get the general idea. Then go through them sentence by sentence. As in the previous exercise, decide how each sentence functions in the passage. If a sentence is out of order, see if you can move it to a more appropriate place.

a. Like visible light, x-rays are a form of electromagnetic radiation. They have a very short wavelength, 0.02 to 10 angstroms. Because of this, they are able to penetrate materials that are normally opaque to visible light. They are also diffracted. These properties make x-rays suitable for direct crystallography. In this process, x-rays are sent into a sample and reflect off the crystal planes. Using known relationships, it is possible to calculate the spacing of these planes. Any deviation from known crystal geometry which is revealed by the diffraction indicates that the material has been strained.

b. The subject of play in animal behavior, probably more than any other subject in animal behavior studies, is open to confusion, misinterpretation and armchair theorizing. Opinions differ even as to its existence and importance. Play, one of the most important activities of young chimpanzees, is the process of acquiring adaptation or organization of behavior in the life span of the individual. It serves two general but important functions. Play provides a behavioral mechanism by which activities appropriate for social functioning can be initiated, integrated, and perfected. The second function of play in social development is to mitigate aggression when it emerges in the young chimpanzee's repertoire. Patterns of social dominance established years earlier may influence his eventual status and status changes. The range and variation in play form and games among chimpanzees is second only to that in humans. Generally, the higher the animal is on the phylogenetic scale, the more frequent and varied is its play.

Chapter Six

FORMAT FOR WRITING

THE PROBLEM

How do you control the presentation of information throughout a whole paper so that the reader can get to the information he or she wants quickly and surely? In what order do you arrange the various parts?

OBJECTIVE

This chapter describes the function of the special formats that many journals require for articles that are submitted to them and many departments require for theses. At the end of this chapter, you will know the purpose of each commonly used section and will have some guidelines for developing your own sections.

JOURNAL FORMAT

Rationale

Technical prose is difficult because it is concentrated. It is not by accident that a reader who reads novels at 750 words a minute may slow down to less than 200 words per minute when confronted with technical prose. The information content of the latter is simply denser and much more specific.

Journals use standard formats largely because technical prose is so difficult, and having a standard format makes it easier for readers and writers alike. Both know exactly how the information will be arranged even before looking at or writing an article.

Journal formats make the reader's (and the writer's) task easier by breaking the whole into distinct parts. Reading may be likened to eating. The more elaborate the dinner, the more important that it be served in separate courses, and that these courses be served in a particular order. This means that you know what to expect and can choose to partake of what parts of the dinner you like. The same is true for the reader—when the paper is broken into clearly defined sections.

Purpose and Content of Sections

Journals differ as to the number of courses served (and their order), but most require some or all of the following.

Title. The title should accurately represent the content of the paper, as the reader often initially decides whether or not to read a paper on the basis of the title alone. Computer literature searches also rely partially on the titles of papers, so titles should be accurate if computer retrieval is to be effective.

Abstract. The abstract also enables the reader to determine whether or not he or she wants to read the entire paper. It should, therefore, be a concise and accurate summary of the entire paper. There are three types of abstracts:

1. *Subject*—states only the subject or hypothesis of the paper
2. *Conclusion*—describes only the conclusions of the paper
3. *Combination*—provides both the subject and the conclusions of the paper

The combination abstract is preferred by most journals. It is also the most useful type.

Introduction. The introduction establishes the background of the paper for the reader. It describes the nature, purpose, and scope of the research and usually includes a brief review of previous work in the field. The introduction prepares the reader for the more complicated information which follows. It's also where the writer establishes her or his competence as a writer and researcher. It's like the first course served in a restaurant. Just as a good soup prepares the diner for a good meal, a good introduction whets the appetite of the reader. A poor one may discourage the reader so much that he or she won't read any further.

Materials and Methods. This section should contain a full description of what was done, how it was done, and the chemicals, tools, equipment, and other experimental materials that were used. It should provide enough detail to enable the reader to evaluate the validity of the conclusions and to repeat the procedure if necessary. Careful notes should be kept of all experimental procedures so that this section can be written accurately and completely.

Results. The results section contains the observations, measurements, and/or other data obtained from the study. This should be a simple section, with as much of the data as possible presented in tables and/or figures.

Discussion. It is in this section of the paper that the author should analyze the results, draw conclusions, make predictions, relate the findings to other work in the field, and suggest areas of future research. This part of the paper will be read more carefully than any other, so it is important that the ideas be well organized and clearly stated.

Using an Established Format

There is one cardinal rule for using this or any other established format. **When you write the text under a specific section heading, stick to the subject which that heading defines.** For example, when writing the **Methods**, that's *all* you describe under that heading. If within this section you use a subheading, "The Johnson and Kumali Method for Culturing Bacteria in Dilute Hydrochloric Acid," that's *all* you describe under this subheading.

This rule is an *absolute* rule because your reader uses the journal format to find the kinds of information he or she is particularly interested in. Someone interested in the method you used to carry out an experiment will go to your Methods section to find this information. If you've put descriptions of your methods into your Results or Discussion, someone trying to repeat your experiment may spend time and money and not be able to duplicate your results. Of course, readers are smart enough to spot incompleteness in methods when they are reasonably familiar with them. Nevertheless, they do not appreciate having to search through your entire paper when all they really want is to copy your method.

Use the journal format, then, to separate your paper into distinct sections, each with its own purpose and content. The information you present will then be more accessible to your readers.

DESIGNING YOUR OWN FORMAT

The standard journal formats are not appropriate for all scientific writing. A scientist may wish, for example, to write something purely speculative—the traditional "armchair" article. Scientists may also find themselves writing such things as environmental impact reports or industrial analyses. Under these circumstances, it is up to the writer to design an appropriate format.

You do this by analyzing the needs of your particular audience, their interests and abilities. You also have to know what you want to tell them. Remember, *the purpose dictates the structure*. Suppose, for example, that you analyze the efficiency of a manufacturing process for a chemical firm. Your audience is executive (management) and not scientific. How do you order your report and divide it into sections which will allow this reader to get at the information which he or she feels is most important? The answer lies in analyzing this person's needs *and* abilities. Such a person is likely to be most interested in your conclusions and recommendations. Unlike the scientist, he or she will *probably not* be interested in your discussion or in the methods you used in your analysis. So, you rearrange your paper to reflect this.

First then, you would define the problem for the manager. Then you would give him or her a summary of your conclusions and recommendations.

Because this person is not likely to have technical expertise, you do this very briefly. You state things in the simplest possible terms. *Then* you write a more or less standard journal article about the subject and attach it to this cover report. This second part is for the people working with the manager, those who are interested in the detail. The information is thus made accessible to the two diverse groups.

Designing a theoretical or speculative discussion is harder, but the same principle applies. You have to know your intended audience, to analyze their interests and background knowledge, and then to determine which portions of your information are most important to them. This will help you decide what kinds of sections to use and in what order.

To illustrate some of the decisions that are involved in this process, the example below has been excerpted from a review paper written by a graduate student. As you read through it, notice how the student has divided it into distinct sections, each with a distinct purpose. We've tried to suggest some of the more important decisions he made while he was doing this. Notice also that, within sections of the paper, the writer has used the basic stimulus-response pattern as a general guide for paragraph structure.

The Origin and Evolution of Maize

Introduction

For approximately one hundred years, maize geneticists have debated the origin of one of the world's primary food souces. For no other major crop has there been such a diversity of evolutionary hypotheses (Beadle 1972).

Because maize has been under artificial selection for thousands of years, it is extremely easy to interpret evolutionary evidence in several ways. Certainly, the present-day maize is of a very "plastic" nature, and it is difficult to relate the evidence collected from a highly cultivated crop to its origin and evolution in the wild condition.

Numerous hypotheses about the origin of maize have been suggested, including:

1. Maize arose from the hybridization of teosinte with an unknown member of the tribe Andropogoneae (Harshbarger 1896, Collins 1912, Kempton 1919);
2. Maize and teosinte have descended along slightly different evolutionary lines from an extinct member of the tribe Andropogoneae (Weatherwax 1918);

How much do you say in an introduction? Here the author has decided to make his introduction brief. Yet he includes some very basic information, *reminding* his reader of what he is sure to have known at one point. In this way he refreshes his reader's memory, leads him or her slowly into the topic, *and* lets his reader know that he knows what he's talking about.

This kind of list is a good way to separate information into bite-sized chunks. Think of how confusing it could be without a listing format.

3. Maize originated from teosinte by artificial selection (by prehistoric people) of the desirable mutations that are characteristic of a cultivated food crop (Ascherson 1880, Emerson and Beadle 1932, Beadle 1939);

4. Modern maize originated from an extinct or undiscovered wild pod maize. Teosinte played no role in the origin of maize. It was, instead a product of maize and *Tripsacum* hybridization (Manglesdorf and Reeves 1939, Manglesdorf 1947, Wilkes 1967).

Evolutionary evidence has allowed the first two hypotheses to be ruled out. However, there is a heated, current debate over the last two—George Beadle and his followers supporting the idea that maize originated from teosinte, Manglesdorf, Reeves, and their students supporting the idea of a wild pod corn origin for maize.

The close relationship among three species *Zea mays* L. (modern maize), *Tripsacum,* and *Zea mexicana* (teosinte) is the basis for this dispute, and the competing hypotheses seek to explain the chronology of their development, each focusing on particular bits of evidence. There is limited evidence to draw upon, but it will now be considered.

Hypothesis 1: Maize Origin from Teosinte

The longest-standing hypothesis proposes that maize is a direct selection from teosinte, or was selected from an ancestor common to maize and teosinte (Ascherson 1880, Emerson and Beadle 1932, Beadle 1939, Beadle 1972). The evolution of maize from teosinte would have come about with prehistoric artificial selection of the four or five major gene differences, many minor genetic differences, and chromosomal mutations that now distinguish maize and teosinte.

The major differences between teosinte and maize are the following: single versus paired pistillate spikelets, two-ranked versus four-ranked phyllotaxy,

This summary is a good way to focus attention on the last two hypotheses.

The writer has moved from summary to a new establishing statement.

The reader's expectations are now firmly established; the writer must discuss the evidence for each hypothesis. Notice that the writer is aware of this and generates a heading which tells the reader that this expectation will be fullfilled.

sessile versus pedicillate pistillate spikelets, and those characters found on chromosome four.

Morphological Evidence for This Hypothesis

The oldest archeological maize from the Tehuacan Valley (Manglesdorf, MacNeish, and Galinat 1967) featured a fragile rachis, allowing for the dispersal of seed, and had long, protective glumes. These are the characteristics which maintain teosinte as a wild plant in Mexico today (Galinat 1971 a).

Ascherson (1880) and Harshbarger (1896) postulated that the polystichous maize ear with its nonseparating rachis arose from the fusion of teosinte's entirely separate spikes. Manglesdorf (1947) concludes that this would have required drastic changes. However, a drastic change may not have been necessary. Many of the floral differences between teosinte and maize are rather simply inherited, and may be due to just four or five major gene differences, along with many minor modifying genes. Beadle (1972) grew 50,000 F_2 progeny of a maize and teosinte hybridization, and was able to recover the parental types in 1/500th of the progeny. This proportion is between the proportions expected if there were four or five gene differences. It must be remembered that maize has gone through as many as ten thousand generations of artificial selection, and it appears that the floral differences separating teosinte from maize are just those characters that would benefit a plant under natural conditions, although Wilkes (1967) has postulated that there is sufficient variability among the six known teosinte races for the derivation of the maize-type ear through natural selection.

Cytogenetic Evidence

Teosinte and maize hybridize very readily in the wild, with teosinte contributing very extensively to total variation of the maize genome. Studies of the pachytene chromosomes of a maize-teosinte hybrid have shown almost complete homology between the maize and

The author has once again adopted the tactic of placing a clear heading before a specific section. He knows, now, that he cannot discuss anything but morphological evidence in this section—and so does the reader. The reader familiar with this evidence can skim this section; the unfamiliar reader will know to read carefully.

Notice here that the writer is closely following the general paragraph format. He introduces a general idea, suggests the conflict, and then presents the evidence, the detail the reader needs to understand the subject. Analysis follows this section quite effectively.

Notice again that the format of this section roughly follows the paragraph and section format suggested in Chapter Five. The first sentence is establishing information; the next three

56

teosinte chromosomes (Ting 1964). Crossing-over between the chromosomes of the hybrid was as complete as that seen in the pure maize genome (Beadle 1939). Pairing is complete between all ten chromosome pairs, except for the short arm of chromosome nine, where no crossing-over was observed between the *C-wx* region (Emerson and Beadle 1932). The lack of crossing-over between this very short chromosome segment was postulated by Emerson and Beadle to be due to a chromosomal inversion in that region.

sections are elaborations, progressively getting more specific; the last is an analysis sentence.

This homologous relationship between maize and teosinte contrasts sharply with the nonhomologous relationship between *Tripsacum* and either maize or teosinte. An F_1 hybrid between *Tripsacum dactyloides* (n = 18) and maize (n = 10) (Manglesdorf and Reeves 1939) showed almost no homology between the chromosomes of maize and *Tripsacum*. Diakinesis figures showed twenty-eight univalents in meiosis. When a maize-tetraploid *Tripsacum* cross was made, eighteen II's (*Tripsacum* chromosomes) and ten I's were seen in the diakinesis figures.

The complex information which follows this establishing statement would be difficult to follow without the framework which is provided by the opening statement.

The paper does not stop at this point, but it is too long to present here in its complete form. The author next presents the evidence for the hypothesis that maize began as a wild pod corn which hybridized with *Tripsacum*. Once this evidence is discussed, the author draws the two hypotheses together in the discussion, pointing out where each is strong and then where each has weaknesses. The conclusion is in the form of a summary and a call for more work, for the author has shown that there is no conclusive evidence for either hypothesis.

For the purposes of this book, the content of the paper is not really important. What is significant is the way in which the author has organized the paper—and the control this organization gives him throughout the discussion. The overall organizing structure used is one of a comparison. The paper follows this outline:

I.   Introduction
II. Hypothesis One
    A. Morphological Evidence
    B. Cytogenetic Evidence
III. Hypothesis Two
    A. Morphological Evidence
    B. Cytogenetic Evidence

IV.   Analysis: Hypothesis One versus Hypothesis Two
       A.   Morphological Evidence (One versus Two)
       B.   Cytogenetic Evidence (One versus Two)

Within this very general *comparison* structure, the author takes pains to develop each idea according to the basic stimulus-response format described in Chapter Five. He also uses headings to tell the reader what the content of each major section will be. This lets the reader know what to expect—and the reader is likely to respond positively when the author fulfills these expectations.

SUMMARY

Established journal formats make both the reader's and the writer's tasks easier by breaking the whole into distinct, manageable parts. Sections commonly used in journal formats include the Title, Abstract, Introduction, Methods, Results, and Discussion.

Whether you're using a predesigned format from a scientific journal or designing your own, there is good reason for breaking your work into distinct sections, each with a specific function in your overall design. Doing this allows you (and the reader) to concentrate on one topic at a time. Once you have established a pattern, *stick to it*. Make sure that you don't inadvertently mix information which ought to be in an introduction, for example, in with your discussion. Mixing of this sort will confuse the reader, and will make it harder for you to write clearly.

The *incremental method* of writing—writing as you carry out your research—is the best way to take advantage of the feedback between writing and thinking. This method assumes that you can organize your work into distinct parts, that each of these parts can be treated as a separate unit, and that all of the parts can be brought together in the end with very little effort. To do this, you have to organize your work and your final paper quite carefully *in the very beginning.* You must completely define your subject and your general approach to the subject. The most important consideration in this defining process is the *purpose* of the work to be done, for this purpose helps you determine the basic organization of your paper.

Once you have defined your purpose, you can begin to actually give your paper structure. You can subdivide your paper into units, each with a specific function. You can design the charts, tables, and graphs which you'll use to present your data. This will, in turn, help you plan your research; it will give you a concrete idea of exactly what kinds of data you have to collect.

As you prepare to write, you next have to consider your reader. Each section of your paper can be divided into the subsections of *establishing statement, detail, analysis, summary,* and *conclusions.* The overall headings for the sections of your paper can be those which a journal defines for you or those which you define yourself, but each can be broken into these functional subsections.

The end result of this process of organizing is that when you sit down to put your ideas into words, you only have to think about one bit of your paper at a time. You can sort through all of the thoughts you have about a subject, analyze them, and decide which part of the organizational structure each idea fits into. Once you have made these decisions, you're ready to begin expressing the ideas in coherent sequences of clear sentences.

1. This short paper does not follow good journal format. Try to repair it. First, establish distinct sections, each with a heading. Then go through the paper sentence by sentence, placing each sentence in its proper section. When you have finished, the paper should read smoothly, with different types of information separated into distinct and logical sections.

IS A WELL-BALANCED BREAKFAST A GOOD WAY TO START THE DAY?

Three problems are often encountered in the early morning—hunger, grogginess, and boredom. It has been suggested by the author's mother that a well-balanced breakfast, consisting of eggs, toast, and orange juice, might alleviate these problems. To test this hypothesis, the author (who suffers from all three morning problems) prepared such a breakfast and ate it. The prepared eggs, toast, and orange juice were taken to a dining table where they were eaten.

Previous experiments of this type have omitted salt and pepper and have used margarine instead of butter. This author found these previous methods of preparation to result in unpalatable meals. The modifications reported in this paper produce a much more tasty meal.

Ingestion of the meal described in this paper was a quite satisfactory solution to the problem of morning hunger. The author's hunger was adequately satisfied. However, the mental alertness of the author remained rather low and boredom set in during the course of the experiment.

One 180 ml can of Citrus Joy Frozen Concentrated Orange Juice was mixed with 540 ml tap water in a Macerette food blender for one minute. 250 ml of the orange juice was poured into a 300 ml drinking glass and refrigerated at 5°C.

Approximately 25 ml Acme Pure Vegetable Salad Oil was placed in a 22 cm cast-iron frying pan and heated to about 150°C on a Goodheet electric range. Two eggs (Grade AA Large) were gently broken open by tapping them on the edge of the frying pan, and their contents were allowed to drop gently onto the heated oil. Any shell fragments which fell into the frying pan were removed and discarded. Table salt and ground black pepper were sprinkled sparingly over the surface of the eggs. When the transparent part of the eggs had become completely opaque white (approximately three minutes), the eggs were gently inverted with a stainless steel spatula and allowed to cook for an additional fifteen seconds, after which they were removed from the frying pan with the spatula and placed on the preheated ceramic plate with the toast.

Two slices of whole wheat bread (Grograin Honey Wheat) were placed in a Burnem toaster and toasted for two minutes. 12 g butter (Grade A Sweet Cream Butter) was spread evenly over each slice. Both slices were then placed on a ceramic plate and kept warm in a Goodheet electric oven at 94°C. It has been reported that freshly brewed coffee can have a significant effect on mental alertness. Perhaps future experiments in this field should include coffee as part of the meal.

All food items used were purchased at a local grocery store. Eggs, butter, and orange juice were maintained at 5 °C prior to use. Bread and oil were kept at room temperature (23 °C).

Although the meal did satisfy hunger, it did not alleviate the problems of lack of mental alertness and breakfast boredom. It is this author's experience that breakfast boredom is frequently alleviated by the comics page of the morning newspaper. Future breakfast experiments in this laboratory will include the newspaper with the meal.

Chapter Seven

STYLE

THE PROBLEM

What is style? What constitutes an effective style for scientific writing? How do you achieve such a style—without years of practice and trial and error?

OBJECTIVES

This chapter takes a look at language in an effort to define the characteristics of various styles. It examines the elements of writing and of styles of writing and explores some of the most common style problems which scientific (and other) writers face. When you have finished the chapter, you should be able to begin to analyze your own style, and to determine whether or not it is appropriate for scientific writing.

What Is Style?

Writing is a communication process. It begins with an idea or a perception of some phenomenon. The writer's task is to transmit the idea or perception *to* the reader. Thus, it isn't the idea that the reader receives, it's the writer's *expression* of this idea. Take, for an example, the following two sentences:

1. The moon ascended to its place above the earth's horizon.
2. As the earth rotated on its axis, the moon came into view and appeared to rise above the horizon.

Each of these sentences is an expression of the same phenomenon, but each suggests an entirely different picture to the reader.

The idea that the reader gets from the first expression is that the moon is rising to some predesignated place above the horizon. The second statement briefly describes the relationships that exist among the moon, the earth, and the viewer, and indicates how they account for the observation. These illustrate two different styles of communicating. Style is, then, the way in which the author *transmits* an idea to a reader, or, *translates* an idea for a reader. The purpose of this translation process is to enable the writer to *share* a particular perception with the reader.

The translator achieves this goal by selecting expressions which have a one-to-one equivalence with the idea that is being expressed. This means that the translator-writer must fully understand the idea in the first place, must understand the elements of the relationship being described *and* the way in which the relationship is structured. The need for complete understanding is the reason the scientific writer should be completely organized before beginning to write—and should write from a complete set of graphics rather than designing graphics after writing. Once the translator-writer feels that the idea is firmly in hand, the next step is to select those terms which precisely represent the elements of the idea and to put them into the right order. The writer has to start to put sentences together.

BUILDING IDEAS WITH SENTENCES

Each time you write, you are translating ideas into sentences. In this process, you make dozens of decisions about how to express each idea, and you make them very quickly. In order to gain control over the way in which you express ideas, you need to understand what is involved in this process of translation.

Each sentence is thus an expression of an idea. And like ideas, sentences start out simply and grow complex as you think more deeply. Sentences begin with what is called a CORE and are expanded by the addition of MODIFIERS. For example:

1. Cell division is triggered. (the CORE idea)
2. Cell division is triggered *by a change in the environment.* (the CORE has been MODIFIED)
3. Cell division is triggered by a change in the environment *such as a change in pH.* (the MODIFIER has been MODIFIED itself)
4. *Possibly,* cell division is triggered by a change in the environment such as a change in pH. (the CORE idea has been MODIFIED again)

This sentence-building process can take place at speeds of over five hundred words per minute. In the process, the writer examines the elements of the idea, analyzes the different ways the core idea can be expressed and the different ways in which it can be modified, and makes decisions. These decisions happen so quickly that one is seldom aware of them. The process is often automatic, performed without conscious thought or effort.

An interesting example of this sentence-building process can be found in ordinary conversation, when two people go through the sentence-building procedure *consciously.* A conversation will begin with a statement or a question. If there is any argument, the two participants will fuss over the wording of the original statement until they're satisfied that they both understand the other's position. Sometimes this semantic exercise can go on for hours. It usually ends when one party says something like "*Now* I understand!" What has happened is that a precise expression of an idea has been built—slowly and carefully. The two have arrived at a *structure* for the idea that is mutually acceptable; the idea itself may not have been modified at all.

Making decisions about how to build a sentence—about how to express the core and then modify this expression—can seem quite complex. But when you begin to analyze the process, you begin to see how basically simple it is. In fact, there are only three *kinds* of decisions which the writer has to make while building sentences. If these kinds of decisions are made consistently and appropriately, the result is a writing style which is effective for communicating with the scientific reader.

THE ELEMENTS OF STYLE

The kinds of decisions you make as you build sentences are:

1. decisions about NAMING
2. decisions about PREDICATING
3. decisions about MODIFYING

Your writing style is a function of the decisions you make about these processes, about what to call things, about how to define the relationships you wish to express, and about how to modify the basic ideas to make them more precise.

Naming

Naming is the most elementary linguistic process. It involves learning a set of symbols (words) which are equivalent to certain things. It is the first stage of learning language, and children learn the rule of "equivalency" quite quickly. They learn, for example, that there is a one-to-one correspondence between spinach and a slightly bitter-tasting green mash that comes in a jar. They learn that the relationship between this word and the green mash is an exclusive one; when they ask for spinach they always get spinach—not candy or apples or meat. It isn't until they get older that they learn that there are dozens of terms which are only roughly equivalent. There are things which can be called by several different words, and there are words which name several different things.

The process of naming sometimes requires the writer to make some fairly difficult decisions because of this overlap in vocabulary. Suppose, for example, you were a manager who had to choose between the terms "carry out," "implement," "operationalize," and "put into effect." Which one should you choose? Novelists and poets choose words to create special effects. They can often enhance the richness of their work by selecting obscure or ambiguous words. Managers (and scientists), on the other hand, usually want to be *understood*. They strive for simple precision in their language. As a competent manager, then, you should pick the **lowest common denominator**, the simplest term which is also precise. The mathematician reduces 222/333 to 2/3; both mean the same thing, but the second is simpler to read. To write clearly, you should simplify, too.

The lowest common *precise* denominator may not always be a simple term. A fisheries biologist in British Columbia may use the term "Tyee salmon" to describe what a biologist in California would call a "King salmon." The lowest *common* denominator may thus *have* to be *Oncorhynchus tshawytshaw.* You might write about this salmon like this.

> *Oncorhynchus tshawytshaw,* the King (or Tyee) salmon, has been studied extensively as part of many state fisheries programs because of its value as a game fish. The King salmon is thus well known as an example of the salmonids. Other varieties which are well known are. . . .

Once you've used the correct term, and then given the reader a simpler term to work with, you can use the simpler term for the rest of the paper as long as there's no chance of misunderstanding. Doing this is generally worthwhile because it expands your audience without forcing you to sacrifice clarity.

Predicating

Predicating is the process of using verbs to draw a relationship between two objects, or simply of stating existence. A child begins to learn to do this at about one year of age. If the child has a name for him/herself and a name for "daddy," *some* communication is possible simply by saying "I, daddy." But when the child learns to express a relationship like "I *want* daddy," more precise communication begins to be possible. This is predicating.

By the time you finish college, you've learned a lot of complex relationships and even more ways of expressing them, of predicating. The problem is to get the process under sure control so that you can be certain that the relationship you perceive and wish to express is the relationship that you *do* express. You want to make this relationship as easy for your readers to see as possible within the limits set by precision.

Examine the following mathematical expressions. Each is an example of a different way of predicating the basic idea "$X = Y$."

$$X = Y$$
$$X - Y = 0$$
X is not greater than nor less than Y
X is not unequal to Y
Equality exists between X and Y

Each of these expressions says exactly the same thing. The mathematician, though, would usually choose "$X = Y$" because it is the simplest expression.

Decisions about predicating in English are a bit more complex than those in mathematics; there are more verbs in English and more structuring options. But the process is essentially the same—the writer chooses an order for the elements of the idea and generates a verb to draw the essential relationship between the elements. For example, examine the following sentences. Each is differently predicated but attempts to express the same idea.

1. Fusion neutrons can be used to convert nonfissionable isotopes of uranium or thorium to fissionable isotopes.

FN
↓
nfi
↓
fi

2. Nonfissionable isotopes of uranium or thorium can be converted by fusion neutrons into fissionable isotopes.

nfi
↓ FN
fi

3. Conversion of nonfissionable isotopes of uranium or thorium to fissionable isotopes is feasible using fusion neutrons.

nfi
↓
fi

(FN)

4. Fusion neutrons may also be used to effect the conversion of nonfissionable isotopes of uranium and thorium to fissionable isotopes.

FN
↓↘
nfi
↓
fi

Although all of these constructions have their place in scientific prose, the simplest constructions (1 or 2) are generally clearest. The point that is important here is that the writer should *choose* the basic structure of the core of the sentence *deliberately*.

Modifying

Modifying processes are the last learned as children, and they are the most difficult to master. Yet controlling them is essential to a good scientific style. This control begins with the knowledge that there are two different kinds of modifiers—simple and complex—and that they are *sometimes* interchangeable. Of course, if they are interchangeable, the simplest form will allow the easiest reading. Compare the following examples:

1. the catalogue *of chemical supplies*   (complex)
   the *chemical supplies* catalogue   (simple)
2. chemical reactions *which rotate*   (complex)
   rotating chemical reactions   (simple)
3. *in addition to this factor*   (complex)
   also   (simple)
4. beams *of laser light that converge*   (complex)
   converging laser beams   (simple)

67

5. a part *of the buffering process*
   which is important
   an *important* part *of the buffering* (simple-
   process   complex)

(complex)

There is nothing wrong with complex modifiers. In fact they can be very useful. It is important for the writer to realize, however, that too many complex modifiers can make a sentence difficult to follow. They get in the way of the CORE elements of the sentence (the subject, verb and object). Take, for example, this description from an administrative manual for secretaries and clerks.

> *Unless any agency obtains an exemption from Accounting Systems Branch, Audits Division, Department of Finance,* all expenditures *of less than $150 that, with respect to an item of nonexpendable property, (a) add to it or (b) improve, better, or as an extraordinary repair and maintenance item extend its originally estimated life,* will be considered to be revenue expenditures *that benefit the current period.*

The abundance of complex modifiers is the problem here, along with the fact that they're placed between the CORE elements. The modification process just ran away with the author. In order to regain control, the author needs to change the form and placement of the modifiers.

> When adding to, improving, or extending the life of nonexpendable property, expenditures of less than $150 will be considered current period expenses, not capital expenses. Obtain exceptions from Accounting Systems Branch, Audits Division, Department of Finance.

Anyone familiar with scientific prose realizes that control of this modification process is not universal among scientific writers either.

CHOOSING A STYLE FOR SCIENTIFIC WRITING

A scientific writer needs to make decisions about Naming, Predicating, and Modifying which will produce a style that is appropriate to the needs of the scientific audience. This ought to be done consciously and methodically. It begins with a consideration of the characteristics of an effective writing style for science.

The Characteristics of a Good Scientific Style

When you're writing for a scientific audience, your primary concern is with your ideas. You have something specific to say and you want to transmit this to your audience as clearly as possible. Your aims as a writer are, then:

1. **Precision.** You don't want to be misunderstood. You want your ideas

to be understood exactly—because scientific ideas are *exact* ideas. When you examine your writing, then, you should check it *first* to make sure it's precise.

2. **Brevity.** You know, from your own reading, how inconvenient it is to have to wade through twenty pages in order to find out what the writer is talking about. Your object in writing is to make sure this doesn't happen to *your* readers.

These two aims are obviously related. If you say something precisely the first time, you don't have to spend paragraph after paragraph saying it over again. The relationship between precision and brevity is an important one. Imprecision generates wordiness which in turn generates further imprecision. Your editing efforts ought to be focused, then, on achieving precision of expression. In the process of doing this you will probably eliminate most sources of wordiness.

Analyzing Your Writing for Precision and Brevity

During the drafting process, you make hundreds of decisions about how to express your ideas. Thinking about being precise and brief while you're writing your first draft is useful, but concentrating too hard on each decision may lead to more problems than it will solve. In drafting, it's most important to get all of your ideas down on paper, in a well-organized fashion, fairly quickly. Then you can go back and edit to achieve a consistently precise and brief style.

The editing process begins with an analysis of your writing. This analysis can be a fairly simple task if you treat your writing methodically. You can begin this process of systematically examining your style simply by applying the principles outlined in this chapter. For example, take a *general* look at the way in which you make decisions about Naming. Count the number of highly specialized terms you have in your writing. If you have a lot of them, you might consider that your reader—however well-versed in the subject she or he may be—may find your work difficult to interpret. You might then consider compensating for the number of specialized terms by trying to simplify modification or predication.

You can analyze and adjust your style in general ways, then, by varying the kinds of decisions you make when you write. The exercises at the end of this chapter are designed to help you practice this kind of analysis. After you've finished them, go on to Chapter Eight; in this chapter you'll find a set of *specific* guidelines for editing.

SUMMARY

When you write, your language is a medium for your thoughts, and you are relying on it to express each thought clearly to the reader. As a scientific writer, your object is to state each idea as accurately and efficiently as possible. Doing this involves building sentences methodically and editing carefully, at all times keeping *Naming, Predicating,* and *Modifying* processes as simple as possible. This requires you to make conscious judgments about your choices of nouns, verbs, and modifiers. These decisions will, of course, become easier the more you write. But you can begin now.

1. Identify the core of each of these sentences. Remember that the core of a sentence is the subject plus the verb (and the object if it is present).

   a. Dwarfness is a desirable feature of the *Begonia semperflorens* complex.
   b. Interest in developing methods for detecting impurities in soybean oil has increased in recent years.
   c. I plan to work on a method to remove the inhibitory substance.
   d. In contrast, the question of differential resource utilization by sympatric species has apparently been little investigated.
   e. These selection pressures have been chosen because they represent real agricultural problems, resistance to which could be agriculturally important.
   f. Those cells whose growth exceeds the mean are retained and subcultured and again subjected to the selection pressure.

2. Simplify each of these unnecessarily complex phrases.

   a. at a high rate of speed
   b. had a stimulatory effect
   c. a pellucid piece of prose
   d. a multiplicity of reasons
   e. the preparation of the sample was accomplished
   f. at this point in time
   g. monitoring of the effluent was effected
   h. phytophageous epigean
   i. under conditions of low temperature

Chapter Eight

EDITING
SCIENTIFIC PAPERS

THE PROBLEM

As you write, you make literally hundreds of decisions about terms and how to put them together to build sentences. You intend the cumulative result of your decisions to be a paper which is relatively easy for the reader to understand. The problem that all writers face lies in making *enough* right decisions to ensure that the end product is readable. Evaluating your own writing, though, is difficult. You become so familiar with your prose that it is hard for you to judge whether or not your efforts to communicate have been successful, whether you have actually achieved precision, brevity, and clarity in your writing. Yet it is as important to be able to evaluate your own writing as it is to make judgments about your own research.

OBJECTIVES

This chapter shows you how you can effectively edit your own work. It describes a set of standards that you can use for evaluating your scientific writing, and a simple procedure for applying these standards. At the end of the chapter, you should be able to take an *objective* look at your writing. You should be able to determine whether your writing is clear and precise, and if it is not, you should be able to make the necessary improvement.

SOURCES OF IMPRECISION IN WRITING

An effective scientific style is fairly easy to achieve, once you understand the sources of imprecision in writing. There are, of course, many such sources, but only the most important are mentioned here.

The Placement of Modifiers

Misplaced modifying phrases or clauses can be confusing. There are two general ways in which modifier placement can make a sentence imprecise. Modifiers between elements of the core can make it difficult for the reader to find and understand the core idea. Also, modifiers that aren't close to what they modify can make it hard for the reader to figure out what is being modified.

73

Modifiers within the core. Most readers are all too familiar with sentences in which the modifiers get in the way of the core idea, sentences which schematically look like:

CmmmmmmmmmmmmmCmmmmC,

where "C's" represent the elements of the core and "m's" represent modifiers. This is the cause for much of the obscurity of legal prose, as is evident in this example.

> *John Smith,* hereinafter known as the party of the first part and delegatee of the estate in question as prescribed under Section IV, paragraph fifteen of the state administrative code governing the transfer of estates, *is* hereby *delegated* the sole authority to represent said estate in court.

The core of this sentence is italicized; everything else is a modifier. These modifiers disrupt the flow of the sentence so badly that it becomes difficult to follow. Here is an example of this kind of problem from the scientific literature.

> The plant *cuticle,* which is composed of cutin, a polymerized, oxidized complex of hydroxy fatty acids, which itself is covered and infiltrated with a complex mixture of lipids, *forms* the *cover* for the outer epidermal cells on all aerial parts of the plant.

By the time you finish reading through all of the definitions which the writer has inserted between the core elements, you may well have forgotten what the subject of the sentence was. This kind of sentence can be clarified by breaking it into several sentences, each with a specific purpose.

> The plant cuticle covers the outer epidermal cells on all aerial parts of the plant. It is composed of cutin, a polymerized, oxidized complex of hydroxy fatty acids that is covered and infiltrated with a complex mixture of lipids.

Here, the first sentence states the core idea simply. The second sentence provides additional information.

Lost modifiers. When a modifier is placed too far away from what it is supposed to modify, it is often too close to something else. For example,

> Male Swiss-Webster mice were used as a source of microsomal enzyme *weighing 18-24 grams.*

Obviously it is the mice, not the microsomal enzyme, that weigh 18-24 grams. Yet the modifier is so far removed from the noun it is modifying that the sentence becomes awkward. The reader has to pause and think about it in order to figure out what the writer is trying to say. In a longer sentence, the problem becomes even more serious.

To facilitate the development of a model, research will be conducted into the following topics, *which* will be used to predict the effect of temperature changes on the behavior of this beetle.

The italicized "which" in this sentence is too far away from "model" and too close to "topics." The reader gets the impression that the topics will be used to predict, but it is the model that will be used to predict.

To facilitate the development of a model which will be used to predict the effect of temperature changes on the behavior of this beetle, research will be conducted into the following topics.

Vague (Imprecise) Terms

The specialized terms which are created in every field of science usually have fairly specific definitions. As the fields evolve, however, the vocabulary necessarily shifts. Some words are discarded, others take on new meanings, and some words acquire several different meanings. These can be troublesome to both the writer and the reader. In the field of genetics, for instance, the definition of the word *gene* (a very basic, simple term about as old as the science itself) has undergone many transmutations. Evolutionists, molecular geneticists, and population geneticists use rather different working definitions for this elementary term. As a result, although geneticists working in these different areas of specialization can usually understand one another, this isn't always the case. And sometimes considerable discussion and give and take is required before an understanding is reached. A writer who is aware of the difficulties generated by certain terms can minimize the problems by defining them briefly at the point where they are first used in a paper. For example,

A gene, a sequence of nucleotides which, when transcribed, will produce a biologically active nucleic acid, . . .

or

A gene, the genetic determinant of a human character, . . .

In addition to avoiding or clarifying specialized scientific terms that are in transition, the scientific writer must also select nonspecialized vocabulary with care. The range of words available for expressing a single thing or event is sometimes mind-boggling. Consider these general terms: remedy, cure, nostrum, physic, tincture, medicant, poultice, elixir, balm, palliative, therapy, treatment. They are approximately equivalent to these more precise terms: antidote, counterpoison, counterirritant, antibody, antiseptic, germicide, emetic, antibiotic. Yet the more precise terms are not equivalent to one another. This is so because they contain *more* information, indicating not only *what* they do (to cure or heal) but *how* they do it. A general rule of thumb for scientific writers is, when in doubt, use the more precise term.

Editing for Imprecision

Look for these sources of imprecision as you edit. A sentence with a clear core relationship, properly placed modifiers, and precise terms is usually clear in all respects. If you can identify and eliminate problems in these three areas, you should be able to correct any major problems in style that you encounter.

When you're editing for style, then, treat your task as a doctor treats a patient. Cure the major diseases, the consistent style problems, first (if you find any) and then cure the minor aches and pains.

Following is a list of specific "symptoms" of imprecision, with some examples of each particular disease. Use this list to measure the basic health of your writing.

COMMON PROBLEMS IN WRITING STYLES

Naming Problems

Unnecessarily difficult words and phrases.

Complex: The size of the sample is *isomorphic* to that of the population.
Simple: The size of the sample is *the same as* that of the population.

Complex: *Crack propagation* in an industrial ceramic (porcelain) is followed by an *acoustic emission.*
Simple: *Cracking* in an industrial ceramic (porcelain) is followed by a *noise.*

Imprecise terms.

Imprecise: Allowing people to camp in this wilderness will produce an *unnatural* strain on the environment.
Precise: Allowing people to camp in this wilderness will produce *a new and potentially harmful* strain on the environment.

Imprecise: As each individual *evolves* in response to environmental changes . . .
Precise: As each individual *adapts* in response to environmental changes. . .

Predication Problems

Invalid core relationships.

Invalid: Environmental factors *mediate* behaviour . . .
Valid: Environmental factors *influence* behaviour . . .

Invalid: Nature *is* change.
Valid: Change *is inherent* in nature.

Unnecessarily complex predicate phrases.

Complex: Alkaline treated foodstuffs *are known to be capable of inducing* pathological changes in animal kidneys.
Simple: Alkaline treated foodstuffs *can induce* pathological changes in animal kidneys.

Complex:	These conditions *have the effect of limiting* the number of experiments which we can carry out.
Simple:	These conditions *limit* the number of experiments which we can carry out.

Core ideas not in the core of the sentence.

Not in core:	*It has been shown* by other workers that in High Temperature Short Time processes energy/cost factors are reduced.
In core:	High Temperature Short Time *processes reduce* energy/cost *factors.*
Not in core:	The *assumption has been made* that this energy loss is related to factors other than convection.
In core:	This energy *loss has been assumed to be related* to factors other than convection.

Core elements separated by too many modifiers.

Separated:	Three *mixtures* of cultures of *Streptococcus thermophilus* to which *Thermobacterium joghurt* had been added and *mixtures* of them and *Leuconostic cremoris* which was produced in raw-milk cream *were analyzed* to determine acidity.
Intact:	Three *mixtures* of *Streptococcus thermophilus* and *Thermobacterium joghurt were analyzed* to determine acidity. The *acidity* of each mixture *was* also *determined* after the addition of *Leuconostic cremoris* (produced in raw-milk cream).
Separated:	All four *types* including those which have evolved from species of the *semperflorens* group in which are found many of the *Begonia* varieties which are commercially exploited as well as those which have evolved from special species hybridizations begun in the nineteenth century, *bear* little *resemblance* to their ancestors.
Intact:	All four *types bear* little *resemblance* to their ancestors. These types include species in the *semperflorens* group (in which are found many commercially exploited *Begonia* varieties) and others which have evolved from species hybridizations begun in the nineteenth century.

Modification Problems

Unnecessarily complex modification.

Complex:	A cell *in which* the sample material was *contained* . . .
Simple:	A cell *containing* the sample material . . .
Complex:	For glasses *with structures similar* to known crystalline forms . . .
Simple:	For glasses *structurally similar* to known crystalline forms . . .
Complex:	*The anomalous* results *of Lee* and *the contrasting* results *of Bell* . . .
Simple:	*Lee's anomalous and Bell's contrasting* results . . .

Modifiers too far from what they modify.

Distant:	Once inside the test tube, the technician could mix the two liquids.
Close:	The technician could mix the two liquids once they were inside the test tube.

Distant: Examine some of the more common samples of this phenomenon which are described as you work in this laboratory situation in the text.

Close: As you work in this laboratory situation, examine some of the more common samples of this phenomenon, which are described in the text.

EDITING YOUR OWN WRITING

A Procedure for Editing

Being aware of the common style problems isn't enough. You must be able to diagnose these problems in your own writing. Most writers have trouble doing this—even when they are familiar with the standards for precise writing. This is because they are so involved in their own ideas that they can't concentrate enough on the expression of those ideas. They read what they *think* they wrote—not what is actually written down on the page.

It isn't difficult, though, if you use a systematic approach. The steps outlined below will help you to recognize problems and to diagnose them accurately. Once this is done, remedying the situation is relatively easy. Use this simple but effective procedure, then, for editing your own writing.

1. **Underline the core elements of every sentence.** Find the subject, the verb, and the object of the sentence. If there is more than one subject, verb, or object, underline all of them.
2. **Look for the core idea.** Make sure that the main idea is expressed in the core of the sentence and that this basic idea makes sense *by itself.*
3. **Check for modifiers between the core elements.** As a general rule, ten or more words between the subject and the verb is too many.
4. **Look for misplaced modifiers.** When you come across a potential problem, draw an arrow from the modifier to what it modifies. A long arrow means that you should rework the sentence.
5. **Check items for precision.** Have you defined the specialized terms? Examine all words representing qualitative judgments. Check to make sure that you've used these correctly, that the reader will understand what *specific* qualities are being summed up in each word.

Using the Procedure

Follow this procedure. Practice it. Above all, don't just read your papers over. Most writers can repair their writing mistakes once they've found them. This simple set of diagnostic procedures will help you become sensitive to the problems which prevent writing from being precise and brief. Following them, you can transform vague feelings about a paragraph's lack of clarity into concrete thoughts about particular problems. For example, read the following paragraph.

When studying the origin and evolution of living organisms, it is most important that characteristics be compared that can give determination of the amount of genetic variation in populations, as well as the obvious need for determination of similarities and differences among living organisms. Until 1959, the only characteristics available to study genetic variation within populations were rare recessive morphological mutants. These were easily visible, but certainly constitute only the slightest fraction of variation within the genome.

On a quick reading, it becomes apparent that the ideas seem fuzzy and complicated. You're not sure precisely what the writer intended, nor is it immediately obvious how you might make the paragraph more precise. Underlining the cores of the sentences, though, reveals some of the specific problems that contribute to the general lack of precision.

The core of the first sentence (*it is important*) does not contain the main idea of the sentence (*characteristics be compared*). The core of the second sentence (*characteristics were mutants*) doesn't make any sense, although you might not realize that if you didn't know something about genetics. The core of the third sentence (*these were visible*) is too imprecise; the reader is left to wonder whether it was the mutants or the characteristics that were visible.

What should you do when you note these problems? In the first sentence, you should reword the sentence so that the core contains the main idea (this should not be hard when you're dealing with your *own* work). In the second sentence, you need to redefine "characteristics." When you do this, the third sentence will be affected because it is an extension of the second. So you wait until the first two are done before you worry about the third. The finished product might look something like this.

In studies of the origin and evolution of living organisms, the characteristics which determine the amount of genetic variation in populations should be examined, along with those which determine the similarities and differences among living organisms. Until 1959, such comparative studies depended upon the availability of easily visible but rare recessive morphological mutations which constitute only a fraction of the variation within the genome.

Depending on the way you have *interpreted* the writer's original intentions, you might choose to rewrite the paragraph differently. The problem of interpretation makes it difficult for a reader to rewrite someone else's prose. But if it is difficult to determine the author's intent, this is an indication that the author has not been precise in the first place. It is exactly this kind of imprecision that you want to eliminate. By examining the paragraph above in a logical and systematic way, you begin to realize just how much interpreting you're doing as a reader—and how much the writer needed to do some systematic editing.

Following the editing procedure doesn't replace thinking; it only gives direction to your thinking. Analyzing your writing is a bit like diagnosing a patient; a step-by-step examination helps you perceive problems and determine what they mean. It gives you a basis for deciding if your work is healthy or not. And if, after you have examined your writing carefully, you find that it meets the general standards for clarity and precision, you have a fairly reliable indication that you are indeed writing precisely.

Once you've finished editing a particular paper or article, you can take one more step. You can look at all of the specific problems you found and see if they add up to a general tendency. If you find, for instance, that you use a lot of very imprecise verbs, this might be a general problem that you could *concentrate on avoiding* when you next write a paper.

The first few times you use this editing procedure, it is likely you will find that one or two sources of imprecision occur regularly in your writing style. Most writers do.

You do not usually need to examine a very long section of your work to discover such general tendencies. Here for example, is a four-sentence abstract. As you read it, underline the cores of the sentences.

> An ecosystem approach for studies designed to assess and predict the various potential impacts upon fisheries and aquatic ecology of those regions facing imminent development is discussed. The ultimate aim of such an integrated, comprehensive environmental assessment is a detailed analysis of the ecosystem as a whole. Emphasis on research is directed toward determining ecosystem interactions, energy transfer and population dynamics. An interdisciplinary approach toward assessing potential environmental impact and assuring environmental quality is considered.

You probably noticed that the writer relies on long strings of complex modifiers, and that the modifiers are frequently placed between the core elements of the sentences. The paragraph can be easily revised by replacing some of the modifiers, as is done below. But identifying the general pattern is what's important for achieving overall improvement in the writing style. This writer should focus attention on the placement of modifiers. Here's the revised abstract:

> An ecosystem approach to structuring environmental impact assessments is discussed, with emphasis on developing concise, clearly defined project objectives founded on sound ecological theory. A detailed analysis of the ecosystem as a whole is the ultimate aim of such an integrated, comprehensive environmental assessment. The emphasis of this research is thus placed on ecosystem interactions, energy transfer, and population dynamics. An interdisciplinary approach to assessing environmental impact and assuring environmental quality is discussed.

Changes in modifier placement led to other stylistic changes as well. These also improve the paragraph. This is what usually happens when writers change their styles in ways suggested by diagnostic analysis of their writing. One major change creates a lot of minor changes which also contribute to the overall clarity and precision of the writing.

SUMMARY

Style is a word used to describe patterns of writing. Each of us develops certain habits that pervade all our writing, and thus, each of us has our own particular style.

An effective style for scientific writing is one characterized by precision and brevity. Unnecessary complexity is avoided. The core of the sentence is also simple, intact, and contains the most important information. Modifiers are simple and are placed close to the words they are intended to modify. Words are selected carefully, to provide the reader with the clearest possible representation of the writer's thoughts.

An effective style can be achieved by the application of methodical editing procedures to your writing. Immediate changes in style of your polished manuscript result from diagnostic analysis and remediation. More significantly, changes in your basic, first-draft style of writing can be achieved if you analyze your present style, identify whatever troublesome tendencies you may have, and then make a conscious effort to avoid them when you write your next first draft. This writing-editing-analyzing-writing loop will produce substantial improvements in your scientific writing style.

Exercises for
Chapter Eight

Using the procedure outlined in this chapter, identify and correct the stylistic problem(s) in each of the following sentences:

1. The development of useful selective media for psychotrophic spoilage bacteria of fresh poultry was discussed.

2. Several workers have been involved in the study of the detection and measurement of genetic variability by electrophoretic methods.

3. The pH of the solution was adjusted to 6.0 with a few drops of 2N NaOH under pH meter control.

4. White wine fermentation is best at 50 to 60 degrees.

5. The transfer of cells which have been growing on the selective medium for several months to new medium will be performed immediately.

6. It could be that the optimum time is different for various species.

7. It is my intention to discuss cyclic nucleotides.

8. A glass column 1 meter long and 2 millimeters in diameter and a solvent system consisting of various proportions of hexane, methanol, acetic acid, and chloroform, delivered at a rate of 60 milliliters per hour at a pressure of 400 pounds per square inch by a pump.

Chapter Nine

PROOFREADING

THE PROBLEM

Readers are a fussy lot. If they can find something to complain about, they usually will. You're probably fussy yourself about the typographical errors you see frequently in newspapers. A typesetter pushes the wrong button and produces a line which looks like this: ''3$Bn 786'' %#*G,P 2P1GFTRQ.'' In response to this kind of simple, nonsubstantive error (especially if it's in the sports page), people frequently write to editors, cancel subscriptions, or grumble about the decline of western civilization.

Exactly the same thing happens if *you* make a simple typographical error, if you spell a few words incorrectly, or if you type 2/1 for 1/2. Your reader may discredit everything you've said after seeing such mistakes. And it is for this reason that you cannot allow mistakes to slip into your final draft. If you do, the reader may suspect that you've made other mistakes—in your research for example.

Proofreading is not a writing problem *per se,* but because it influences the reader's response to your writing, it's important that you learn to proofread effectively.

OBJECTIVES

At the end of this brief chapter you will have a set of effective procedures that will make it easier for you to find minor errors in your writing or in your typed manuscript—if it contains any mistakes.

PROOFREADING

Proofreading isn't really reading. In fact, the more you *read* while you're doing it, the more likely you are to miss errors. You get involved in the ideas too easily when you read. What you're interested in when proofreading is *perceiving* things like transposed letters, misspellings, commas out of place, E's for W's and so forth. The less you have to think about content at this stage, the better.

Sharpening Your Perception

But when you're reading your own work it is extremely difficult to catch errors, because you are still interested in the development of your ideas. You need some simple ''tricks'' to help you improve your perception of what's actually on the paper. Here are some methods suggested by professional proof-readers.

1. **Hire a professional proofreader.** You might expect this from proof-readers. It's not bad advice, but it could be expensive.

2. **Wait.** Put the paper away for a few days or a few weeks until you've forgotten its content. This can help, but do you have the time?

3. **Use a proofreader's card.** This is a white cardboard card with a thin (4mm) slit in it. It keeps you from skipping from line to line. When you use it, read backwards (from the bottom of the page to the top). Here's what one looks like. You can make one in a few moments:

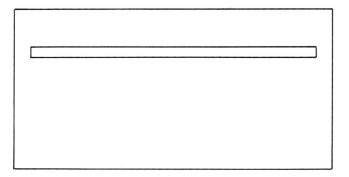

Figure 9.1. Proofreader's card.

4. **Take a break every hour.** You can become so used to your own writing that you don't even notice *major* mistakes. Reading a page or two from a good novel, a comic book, *The National Lampoon*—anything totally non-scientific—can break your reading pattern and thus help you perceive mistakes. Do not read these things for more than about ten minutes during every hour of proofreading. Your object is to proofread, not to give yourself an excuse not to work.

5. **Read aloud, in full voice.** If you find yourself stumbling, stop and look at this passage carefully.

6. **Mix up the pages of your paper.** This will keep you from concentrating on content. Be sure that they are numbered before you shuffle them.

7. **Do your own typing**—if you can type reasonably quickly. Typing forces you to "feel" the letters as you write. It can help you avoid errors in the first place, IF YOU TYPE WELL.

Sharpening Your Knowledge

To proofread successfully you must, of course, know an error when you see one. If you can't spell, don't trust yourself to perceive spelling errors. If your math is weak, don't trust yourself to proofread equations. If you don't know a comma from a semicolon, find someone who does. There is, of course, no excuse for not knowing these things. And if you find that you do not, you might do something about it—quickly. But that is beyond the scope of this book. There are many books on spelling, grammar, and scientific style. Ask a senior colleague in your field which of these are appropriate. In the meantime find somone competent to help you proofread.

1. Proofread the following passage. Circle or underline all errors.

Allthough very little work has beem done on the control of the cell cycle in higher plants. information from cell cycle stidies in animal systems can be applied to plant systems as the regalutory mechanisms are most likely similar

That there are certian control points in the cell cycle is beyond dispute. This has best been demonstrated in *Tetrahymena* where cell cucle progression can be pervented by heat shock oly up to a certain point—the "transition point—after which heat shock has no affect. This sort of critical point has also been de onstrated in a number of other systems, including mammallian cells.

Numerous models have been proposed for control of the cellcycle. According to the most well-developed theory, cell division requires requires the assembly of some sturcture from a number of componants. At least two are protiens and one must be synthesized continouosly is the assembly is to go forward. This structure will disintigrate if the supply of components is interuppted. When complete, however, the structure is stable and it them preforms some essential function in cell division. Temperature shock interupts assembly of the strucutre; it breaks down; and assembly must begin all over agin.

Other models involve sequences of events which must be completed dbefore the cell can divide. These events might be morpological changes or protein synthesis or synthesis or RNA messages.

Other models involve sequences of events which must be completed before the cell can divide. These events might be morpological changes or protein synthesis or synthesis of RNA messages.

Cell division may be triggerred by a change in the envirnment. Punturing a monolayer of mammalian calls in culture stimulates cell division, as does wounding a plant. Treating unfertilized sea urchin eggs (which ordinarily would not divide until fertilized with ammonium hydrozide will stimulate them to divide. This treatment makes the interior of the egg more alkeline, So perhaps Ph changes are involved in cell cycle control

The cell cycle is most likely directed by a programmd sequence of gene activity resulting inthe synthesis of variuos enzymes involved in mitosis and DNA synthesis. That physical changes can trigger cell division may be explained by enzyme activation. In a sea urchin egg, For example, the cycle may have stopped at a certain point with the synthesis of a certain enzyme. When the pH of the sytoplasm is raised with ammonium hydroxide, the enzyme is activated and the cycle progresses by the induction of synthesis of subsequent enzymes.

Chapter Ten

WRITING SYSTEMATICALLY

Each chapter in this book has treated a distinct part of the long writing process which begins when you first sit down to define the subject of your research and ends when you proofread your final manuscript. Each of these steps requires that the writer make some decisions. Writing systematically means that the writer should take certain steps and make certain decisions in *specific order,* and that each step should be evaluated and, where necessary, *corrected* before the next step is undertaken. Guidelines have been provided to help you in making many of the decisions and procedures have been suggested to help you evaluate your decisions.

In this chapter, this whole process is summarized in the form of a flow chart. The chart can help you treat writing as an incremental and systematic process by reminding you of the individual steps to be taken. When you write, use this chart as a guide. Adjust it to suit your own needs. Hang it over your writing area. Use it to remind yourself that writing is a *process,* one that should be started early and be intimately related to your research. This way, you can benefit from the feedback between thinking and writing.

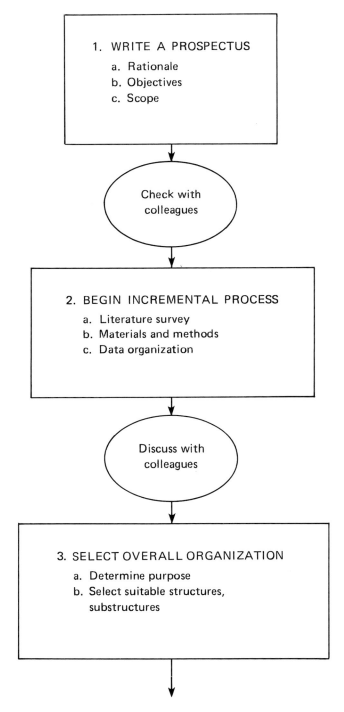

Figure 10.1. Summary flow chart.

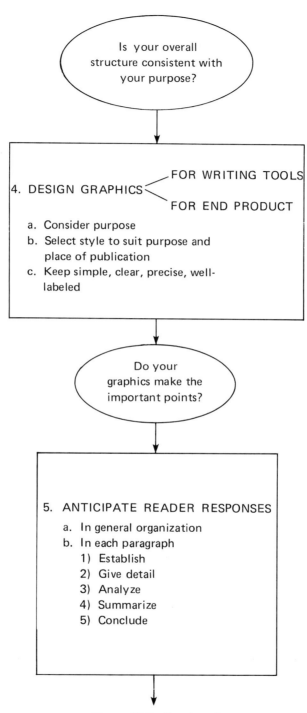

Figure 10.1. Continued.

93

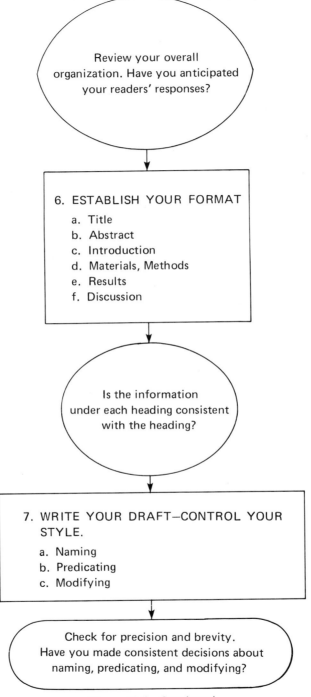

Review your overall
organization. Have you anticipated
your readers' responses?

6. ESTABLISH YOUR FORMAT
    a. Title
    b. Abstract
    c. Introduction
    d. Materials, Methods
    e. Results
    f. Discussion

Is the information
under each heading consistent
with the heading?

7. WRITE YOUR DRAFT—CONTROL YOUR
STYLE.
    a. Naming
    b. Predicating
    c. Modifying

Check for precision and brevity.
Have you made consistent decisions about
naming, predicating, and modifying?

Figure 10.1. Continued.

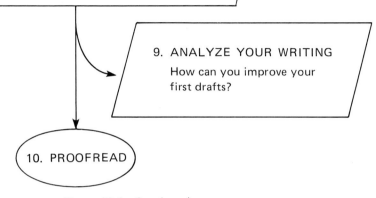

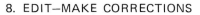

8. EDIT—MAKE CORRECTIONS
   a. Underline cores
   b. Look for core ideas
   c. Check for modifiers between core elements
   d. Look for misplaced modifiers
   e. Check items for precision and brevity

9. ANALYZE YOUR WRITING

   How can you improve your
   first drafts?

10. PROOFREAD

Figure 10.1. Continued.

GLOSSARY

Very few specialized terms are used in this book. You need not memorize a lot of grammar vocabulary. The few terms which we have had to use to describe certain concepts are defined below.

Structure: The ordering of the elements of an idea in sentences, paragraphs, or parts of a paper. The difference between "$2H_2O \xrightarrow{e} 2H_2 + O_2$" and "$2H_2 + O_2 \xrightarrow{\Delta} 2H_2O$" is a structural difference. The elements of the idea are the same, but the relationship between them is different and they are thus differently *structured*.

Incremental Writing: The process of writing a paper or book in brief segments—from an overall plan to the finished product.

Prospectus: The first part of the incremental writing process, the prospectus is a formal summary of a proposed research or writing project.

Sentence: A sentence is a symbolic representation of a relationship. There are two basic kinds of sentences: (1) sentences which say that something exists, and (2) sentences which say that something does something.

Core: The essential elements of a sentence. These elements are the *subject* and the *verb* or the *subject, verb,* and *object.* In the sentence *"Aflatoxins and* their animal biotransformation *products were assayed* for carcinogenic potential," "Aflatoxins and products were assayed" is the core.

Subject: One of the parts of the core. In the sentence "X = Y," "X" is the subject, the thing being defined by "Y." In the sentence "X determines Y," "X" is again the subject, the thing which is acting.

Verb: A word or words which draws a relationship. In the sentence "X = Y," the equal sign is the verb, the element which draws the relationship between the other elements.

Object: This is the "Y" element in either "X = Y" or "X determines Y" or "X does Y" relationships.

Modifier: Any word or combination of words which adds to or makes the core relationship more precise. In the sentence "Aflatoxins and *their biotransformation* products were assayed *for carcinogenic potential,"* all words in italics are modifiers.

Writing: The process of transmitting an idea to an audience by representing it with symbols. This involves three steps: (1) choosing symbols for each element of the idea, (2) putting these together in some order, and (3) modifying the elements with other symbols.

Editing: The process of examining a written translation of an idea to ensure that a reader will receive a precise impression of the original idea.

Proofreading: The process of checking a piece of writing to ensure that there are no minor errors to distract a reader from the ideas.

Spelling: Making words (large symbols) out of letters (small symbols). Different letters, or the same letters in different order, make different words. Remember this.

Answers

SOLUTIONS TO CHAPTER TWO EXERCISES

1. Here is an example which illustrates the way you can break a problem down into its component parts. You can apply these basic steps to the topic of your choice.

Topic: Sand-kicking on public beaches.
Specific Problem: Why does Muscle Mike kick sand in Wimpy Willie's face?

Questions to Be Answered

1. Does Muscle Mike really kick sand in Wimpy Willie's face?
   a. Procedure:
      (1) observe MM and WW at the beach from several angles
      (2) photograph suspected cases of sand-kicking to confirm observations
   b. If no cases of WW having sand kicked in his face by MM are actually observed, change *Specific Problem* to, "Is Wimpy Willie paranoid, and, if so, why?"
   c. If cases of sand-kicking are observed, proceed with the following questions:
2. Is WW a 90-pound weakling?
   a. Procedure:
      (1) weigh WW
      (2) ask WW to perform several tasks requiring average strength
3. Does WW provoke the sand-kicking attacks?
   a. Procedure:
      (1) observe WW's behavior
         (a) words: Does WW call MM nasty names or make rude remarks to him?
         (b) actions: Does WW direct socially unacceptable gestures toward MM?
4. Is MM abnormally hostile?
   a. Procedure:
      (1) break into MM's psychiatrist's office and examine MM's records for evidence of hostility.
5. Does WW just stumble into the path of MM's exuberant, but otherwise harmless, sand-kicking?
   a. Procedure:
      (1) take WW to local medical school and have his locomotor coordination tested

6. If all of the above procedures fail to reveal the answer to the problem, approach Muscle Mike (if you dare) and say to him, "Why do you kick sand in Wimpy Willie's face?"

SOLUTIONS TO CHAPTER THREE EXERCISES

1. Here are some examples of the ways you might organize the fifteen animals. Each represents a different way of looking at data. Although there is no one "correct" organizational pattern for a specific purpose, note how defining your purpose limits the ways you can organize the data.

   a. Purpose: to compare the size of animals

Bigger than Breadbox		Smaller than Breadbox	About Same Size as Breadbox
dog (most)	horse	trout (most)	chicken
lion	elk	goldfish	cat
turkey	pelican	bat	rabbit
elephant	ostrich	parakeet	

   b. Purpose: classification of body covering

Hair		Feathers	Scales
cat	elephant	chicken	trout
dog	bat	turkey	goldfish
rabbit	horse	parakeet	
lion	elk	pelican	

   c. Purpose: to study the relationships between Americans and animals

	Often Kept as Pets	Not Often Kept as Pets
Commonly Eaten by Americans	chicken rabbit	turkey trout
Rarely or Never Eaten by Americans	cat dog goldfish horse parakeet	bat elephant elk lion ostrich pelican

100

2. a. The discrepancy between the simple chronological diagram of events and the subsequent paragraph is obvious here. Whereas the diagram illustrates the sequence of events as

Harvest→Shipping→Weighing→Storage→Testing

the paragraph describes the events in this order:

Shipping→Weighing→Testing→Shipping→Harvest→Storage

Is this discrepancy important? Yes, it is. The paragraph is readable as it stands and, with some effort, you can figure out what is going on. But it would be so much clearer and *easier* to read if it were in chronological order. With the flow-chart as a guide, the flaws in the paragraph are easy to spot and correct.

b. This example is harder to interpret than the first and it is perhaps more typical of the organizational problems you may encounter in your writing. The diagram illustrates the symbiotic relationship between plants and animals and the paragraph begins along this line. Notice, however, that the rest of the paragraph deals only with the benefits which animals derive from plants. The relationship being described is no longer the relationship depicted in the diagram.

Getting off the track, changing the substance of a discussion in midstream, is one of the nost difficult organizational problems to solve. The shift in emphasis may be so subtle as to be almost undetectable. Comparing the paragraph to the diagram, however, forces you to analyze the relationship being presented.

SOLUTIONS TO CHAPTER FOUR EXERCISES

TABLE 4.2

Respiration Rates of Common Vegetables at 0-30°C

Vegetable	Ml CO_2 Produced per kg-hr.			
	$0°$	$10°$	$20°$	$30°$
Broccoli	11	22	89	136
Lettuce	4	12	23	42
Tomato	5	8	20	37
Potato	4	5	6	29
Onion	4	5	6	9

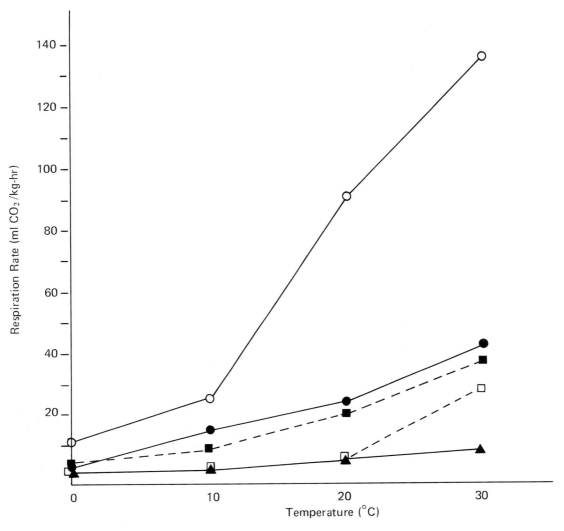

Figure 4.3. Average respiration rates of vegetables at 0-30°C. Broccoli (○), lettuce (●), tomato (■), potato (□), onion (▲). By the colorimetric method.

Figure 4.4. Deterioration of vegetables with increasing temperature (0-30°C).

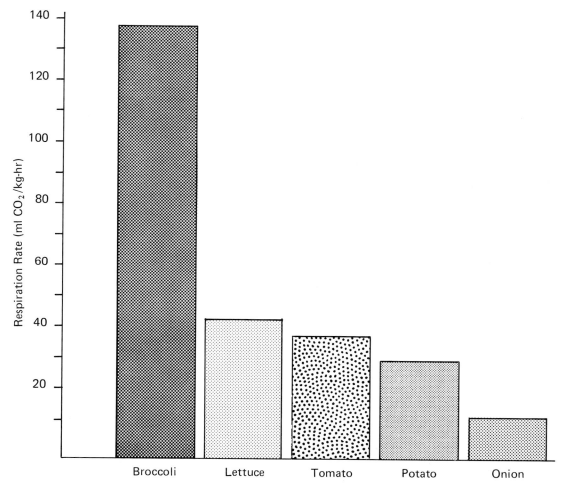

Figure 4.5. Respiration rate at 30°C.

Your figures may not look exactly like ours, but they ought to be similar. Of the four, the diagram gives the reader the least amount of information. Yet this might be all that was necessary if you were giving a talk to a group of grocery store managers. The bar graph gives slightly more information, in that it shows not only the relative order of these vegetables with regard to deterioration but also the magnitude of the differences at one temperature. We chose 30°C for our bar graph because this is where the differences are most dramatic. If 20° C were the prevailing room temperature, however, this would have been the better choice. The table and line graph show data points. In general, one of these formats would be preferred in a scientific paper. The information is more readily discerned in the graph than in the table.

SOLUTIONS TO CHAPTER FIVE EXERCISES

1. For each statement, the *probable* reader response is given in terms of the questions raised by the statement. A good writer will anticipate these questions and make sure they are not left unanswered.

   a. What is "much progress?" Exactly what progress has been made and by whom?

   b. What are the "subtle morphological differences" between the two species?

   c. Why has the use of DES been questioned? Who has questioned it?

   d. What is the novel method?

   e. Who says the hormone enhances brain development? What does "enhance" mean?

2. Here is one way the sets of statements might be organized (there are certainly others).

   I. The presence of a strong chemical or electrical gradient across a cell is called cell polarity. It is especially evident in eggs and zygotes. — Establishing Information

   Schulz and Jensen (1968) have shown that the highly polar *Capsella* egg becomes more polar when fertilized and that the polarity determines the plane of the first cell division. Torrey and Galun (1970) have demonstrated the importance of cell polarity in *Fucus*. When the polarity of a *Fucus* zygote is disturbed by a high sucrose concentration, abnormal embryos develop. — Detail

   Cell polarity is undoubtedly of critical importance in differentiation. — Conclusion

   As this paragraph is merely reviewing existing literature, there are no analysis statements. There is also not enough information presented to warrant a summary.

   II. One of the most important characteristics of the cave environment is the absence of light. — Establishing Statement

   In epigeans (surface-dwellers), light plays an important role in the regulation of the endocrine system. For example, increasing daylength often stimulates reproductive activity. — Detail

   In the absence of light, this regulation cannot occur and cavernicoles (cave-dwellers) consequently have altered endocrine systems. — Analysis

   Caves are devoid of green plants, as photosynthesis cannot take place in darkness. — Detail

Strictly phytophageous (plant-eating) animals are excluded from the subterranean world, and those animals remaining must turn to other food sources.	Transition
There are two primary sources of food energy in the cave, both of which originate on the surface.	Establishing Statement for new paragraph dealing with food sources

3. There are no hard and fast answers to the question "Has the writer satisfied the reader's expectations?" For each passage, one possible interpretation is presented. As you compare these interpretations to yours, think about the way in which making decisions about the function of a sentence in a paragraph forces you to analyze in an orderly fashion.

a. Like visible light, x-rays are a form of electromagnetic radiation.	Establishing Statement
They have a very short wavelength, .02 to 10 angstroms. Because of this, they are able to penetrate materials that are normally opaque to visible light. They are also diffracted.	Detail
These properties make x-rays suitable for direct crystallography.	Analysis and Transition (New Establishing Statement)
In this process, x-rays are sent into a sample and reflect off the crystal planes. Using known relationships, it is possible to calculate the spacing of these planes.	Detail
Any deviation from known crystal geometry which is revealed by the diffraction indicates that the material has been strained.	Analysis

This writer has done a good job in using the general format to guide his or her work. The subject is introduced in very simple form and followed by enough detail to set the stage for the new establishing statement. There is little analysis and no conclusion because the writer wants to discuss crystallography. The whole passage is composed of two sections. Each section presents information in the right order, although not all of the kinds of information in the format are included. At the end of the passage there is a piece of analytical information which also acts to prepare us for the eventual discussion of crystallography.

b. The subject of play in animal behavior, probably more than any other subject in animal behavior	Establishing Statement: Controversy

studies, is open to confusion, misinterpretation, and armchair theorizing.

Opinions differ even as to its existence and importance.	Detail
Play, one of the most important activities of young chimpanzees, is the process of acquiring adaptation or organization of behavior in the life span of the individual.	Establishing Statement: Definition
It serves two general but important functions.	Establishing Statement: Functions
Play provides a behavioral mechanism by which activities appropriate for social functioning can be initiated, integrated, and perfected. The second function of play in social development is to mitigate aggression when it emerges in the young chimpanzee's repertoire. Patterns of social dominance established years earlier may influence his eventual status and status changes.	Detail
The range and variation in play form and games among chimpanzees is second only to that among humans.	Establishing Statement: Comparison with Human Play
Generally, the higher the animal is on the phylogenetic scale, the more frequent and varied is its play.	Detail

When one establishing statement follows another (and here there are four new subjects introduced in the space of less than 150 words), something is wrong. What's wrong here is that the writer has a complex subject and much to introduce to the reader. But he or she has not carefully worked out the relationships between these related pieces of information. The controversy is established even before we know what the subject is, and then the controversy is dropped. What the writer needs to do is rethink the relationships among the four subjects introduced here: animal play, the controversy over animal play, the functions of play in chimpanzees, and the comparison between chimpanzee play and human play. Thinking about these subjects in terms of their appropriate function within the general format, the writer might decide that the controversy and the human-chimp comparison are not really appropriate for the beginning of a paragraph or section of a paper and would be better suited to the discussion section of the paper. Leaving these out, the writer would then be able to define play and discuss its functions in a more orderly fashion.

SOLUTION TO CHAPTER SIX EXERCISE

1. Here is one way in which the paper might be organized. Note that each section is distinct and that no information is placed in an inappropriate section.

IS A WELL-BALANCED BREAKFAST A GOOD WAY TO START THE DAY?

A. *Introduction*

Three problems are often encountered in the early morning— hunger, grogginess, and boredom. It has been suggested by the author's mother that a well-balanced breakfast, consisting of eggs, toast, and orange juice might alleviate these problems. To test this hypothesis, the author (who suffers from all three morning problems) prepared such a breakfast and ate it.

B. *Materials and Methods*

All foods items used were purchased at a local grocery store. Eggs, butter, and orange juice were maintained at 5° C prior to use. Bread and oil were kept at room temperature (23° C).

One 180 ml can of Citrus Joy Frozen Concentrated Orange Juice was mixed with 540 ml tap water in a Macerette food blender for 1 minute. 250 ml of the orange juice was poured into a 300 ml drinking glass and refrigerated at 5° C.

Two slices of whole wheat bread (Grograin Honey Wheat) were placed in a Burnem toaster and toasted for 2 minutes. 12 g butter (Grade A Sweet Cream Butter) was spread evenly over each slice. Both slices were then placed on a ceramic plate and kept warm in a Goodheet electric oven at 94° C.

Approximately 25 ml Acme Pure Vegetable Salad Oil was placed in a 22 cm cast-iron frying pan and heated to about 150° C on a Goodheet electric range. Two eggs (Grade AA Large) were gently broken open by tapping them on the edge of the frying pan, and their contents were allowed to drop gently onto the heated oil. Any shell fragments which fell into the frying pan were removed and discarded. Table salt and ground black pepper were sprinkled sparingly over the surface of the eggs. When the transparent part of the eggs had become completely opaque white (approximately 3 minutes), the eggs were gently inverted with a stainless steel spatula and allowed to cook for an additional 15 seconds, after which they were removed from the frying pan with the spatula and placed on the pre-heated ceramic plate with the toast.

The prepared eggs, toast, and orange juice were taken to a dining table where they were eaten.

C. *Results*

The author's hunger was adequately satisfied. However, the mental alertness of the author remained rather low and boredom set in during the course of the experiment.

D. *Discussion and Conclusions*

Ingestion of the meal described in this paper was a quite satisfactory solution to the problem of morning hunger. Previous experiments of this type have omitted salt and pepper and have used margarine instead of butter. This author found these previous methods of preparation to result in unpalatable meals. The modifications reported in this paper produce a much more tasty meal.

Although the meal did satisfy hunger, it did not alleviate the problems of lack of mental alertness and breakfast boredom. It has been reported that freshly brewed coffee can have a significant effect on mental alertness. Perhaps future experiments in this field should include coffee as part of the meal. It is this author's experience that breakfast boredom is frequently alleviated by the comics page of the morning newspaper. Future breakfast experiments in this laboratory will include the newspaper with the meal.

SOLUTIONS TO CHAPTER SEVEN EXERCISES

1. The core of each sentence is in bold face.

   a. **Dwarfness is** a desirable **feature** of the *Begonia semperflorens* complex.

   b. **Interest** in developing methods for detecting impurities in soybean oil **has increased** in recent years.

   c. **I plan to work** on a method to remove the inhibitory substance.

   d. In contrast, the **question** of differential resource utilization by sympatric species **has** apparently **been** little **investigated.**

   e. These selection **pressures have been chosen** because they represent real agricultural problems, resistance to which could be agriculturally important.

   f. Those **cells** whose growth exceeds the mean **are retained** and **subcultured** and again **subjected** to the selection pressure.

2.

Complex	Simple
a. at a high rate of speed	quickly, fast
b. had a stimulatory effect	stimulated
c. a pellucid piece of prose	clear writing
d. a multiplicity of reasons	many reasons
e. the preparation of the sample was accomplished	the sample was prepared
f. at this point in time	now
g. monitoring of the effluent was effected	the effluent was monitored
h. phytophageous epigean	plant-eating surface- dweller
i. under conditions of low temperature	at low temperatures

SOLUTIONS TO CHAPTER EIGHT EXERCISES

Although only one improved version of each sentence is presented here, there are certainly other ways to improve the sentences. The core of each sentence is in bold face.

1. The **development** of useful selective media for psychotrophic spoilage bacteria of fresh poultry **was discussed.**

   Problem: Core elements separated by too many modifiers.

   Solution:
   (Subject) discussed the **development** of useful selective media for psychotrophic spoilage bacteria of fresh poultry.

2. Several **workers have been involved** in the study of the detection and measurement of genetic variability by electrophoretic methods.

   Problem: Unnecessarily complex predication. "Involved" and "study" are unnecessary in this sentence. It is the "detection" and "measurement" which are important.

   Solution:
   Several **workers have detected** and **measured** genetic variability using electrophoretic methods.

3. The **pH** of the solution **was adjusted** to 6.0 with a few drops of 2N NaOH under pH meter control.

   Problem: Complex modification. As most of the information in the modifying phrases is unnecessary, it can just be eliminated rather than simplified.

   Solution:
   The **pH was adjusted** to 6.0 with 2N NaOH.

4. White wine **fermentation is best** at 50 to 60 degrees.

   Problem: Imprecise sentence. What does "best" mean in this case? Is the temperature in Fahrenheit or Celcius degrees? We must guess at the writer's exact meaning to rewrite the sentence.

   Solution:
   White **wine is** most **flavorful** when it has been fermented at 50 to 60 degrees Fahrenheit.

5. The **transfer** of cells which have been growing on the selective medium for several months to new medium **will be performed** immediately.

   Problem: Core elements separated by too many modifiers.

   Solution:
   I will immediately **transfer** the **cells** which have been growing on the selective medium for several months to new medium.

6. **It could be** that the optimum time is different for various species.

   Problem: Core idea not in the core of the sentence.

Solution:
The optimum **time may differ** for various species.

7. **It is** my **intention** to discuss cyclic nucleotides.

   Problem: Unnecessarily complex predication. The core idea is also not in the core of the sentence.

   Solution:
   I will discuss cyclic **nucleotides.**

8. A glass **column** 1 meter long and 2 millimeters in diameter and a solvent **system** consisting of various proportions of hexane, methanol, acetic acid, and chloroform, delivered at a rate of 60 milliliters per hour at a pressure of 400 pounds per square inch by a pump.

   Problem: This sentence does not have a complete core! There are two subjects, here, "column" and "system," but no verb. Everything else is modification. When you start stringing together so many modifying phrases, it is very easy to forget what the core idea is.

   Solution:
   A glass **column** (1 m long and 2 mm diameter) and a solvent **system were used** to elute the compounds. The solvent **system consisted** of hexane, methanol, acetic acid, and chloroform and **was pumped** at 60 ml per hour at a pressure of 400 lb per square inch.

SOLUTIONS TO CHAPTER NINE EXERCISES

1.   Although very little work has been done on the control of the cell cycle in higher plants, information from cell cycle studies in animal systems can be applied to plant systems as the regulatory mechanisms are most likely similar.

   That there are certain control points in the cell cycle is beyond dispute. This has best been demonstrated in *Tetrahymena* where cell cycle progression can be prevented by heat shock only up to a certain point—the "transition point"—after which heat shock has no effect. This sort of critical point has also been demonstrated in a number of other systems, including mammalian cells.

   Numerous models have been proposed for control of the cell cycle. According to the most well-developed theory, cell division requires the assembly of some structure from a number of components. At least two are proteins and one must be synthesized continuously if the assembly is to go forward. This structure will disintegrate if the supply of components is interrupted. When complete, however, the structure is stable and it then performs some essential function in cell division. Temperature shock interrupts assembly of the structure; it breaks down; and assembly must begin all over again.

   Other models involve sequences of events which must be completed before the cell can divide. These events might be morphological changes or protein synthesis or synthesis of RNA messages.

Cell division may be triggered by a change in the environment. Puncturing a monolayer of mammalian cells in culture stimulates cell division, as does wounding a plant. Treating unfertilized sea urchin eggs (which ordinarily would not divide until fertilized) with ammonium hydroxide will stimulate them to divide. This treatment makes the interior of the egg more alkaline, so perhaps pH changes are involved in cell cycle control.

The cell cycle is most likely directed by a programmed sequence of gene activity resulting in the synthesis of various enzymes involved in mitosis and DNA synthesis. That physical changes can trigger cell division may be explained by enzyme activation. In a sea urchin egg, for example, the cycle may have stopped at a certain point with the synthesis of a certain enzyme. When the pH of the cytoplasm is raised with ammonium hydroxide, the enzyme is activated and the cycle progresses by the induction of synthesis of subsequent enzymes.